# CONTEMPORARY ISSUES IN REGIONAL PLANNING

# Contemporary Issues in Regional Planning

*Edited by*

TIM MARSHALL
*Oxford Brookes University*

JOHN GLASSON
*Oxford Brookes University*

PETER HEADICAR
*Oxford Brookes University*

LONDON AND NEW YORK

First published 2002 by Ashgate Publishing

Reissued 2018 by Routledge
2 Park Square, Milton Park, Abingdon, Oxon, OX14 4RN
711 Third Avenue, New York, NY 10017, USA

*Routledge is an imprint of the Taylor & Francis Group, an informa business*

Copyright © Tim Marshall, John Glasson and Peter Headicar 2002

All rights reserved. No part of this book may be reprinted or reproduced or utilised in any form or by any electronic, mechanical, or other means, now known or hereafter invented, including photocopying and recording, or in any information storage or retrieval system, without permission in writing from the publishers.

Notice:
Product or corporate names may be trademarks or registered trademarks, and are used only for identification and explanation without intent to infringe.

Publisher's Note
The publisher has gone to great lengths to ensure the quality of this reprint but points out that some imperfections in the original copies may be apparent.

Disclaimer
The publisher has made every effort to trace copyright holders and welcomes correspondence from those they have been unable to contact.

A Library of Congress record exists under LC control number: 2001095434

ISBN 13: 978-1-138-72400-6 (hbk)
ISBN 13: 978-1-138-72395-5 (pbk)
ISBN 13: 978-1-315-19271-0 (ebk)

# Contents

List of Tables, Figures and Boxes — vii
List of Contributors — ix
Acknowledgements — xi
List of Abbreviations — xii

**PART I: INTRODUCTION** — 1

1. Introduction — 3
   *Tim Marshall*

**PART II: REGIONAL REALITIES AND REGIONAL INSTITUTIONS** — 21

2. Thinking about the English Regions — 23
   *Allan Cochrane*

3. Scenarios for the Future of Regional Planning Within UK/EU Spatial Planning — 35
   *Jeremy Alden*

4. The Regional Agenda, Planning and Development in Scotland — 54
   *Greg Lloyd*

5. Government Offices for the Regions and Regional Planning — 70
   *Mark Baker*

6. Integrating Local and Central Concerns in the New Regional Landscape: Pipe Dream or Political Project? — 88
   *Martin Stott*

## PART III: THEMES AND TOPICS — 95

7 Regional Transport Strategies: Fond Hope or Serious Planning? — 97
*Peter Headicar*

8 Re-Shaping Regions, Re-Ordering Utilities: Competing Development Trajectories — 110
*Simon Marvin*

9 Promoting Sustainable Development through Regional Plans: The Role of the Environment Agency — 125
*Hugh R Howes*

10 Regional Planning and the Environmental Dimension of Sustainable Development — 145
*Peter Roberts*

## PART IV: REGIONAL PLANNING – REGIONAL VARIATION — 163

11 Integrated Policy Development at the Regional Level – A Case Study of the East Midlands Integrated Regional Strategy — 165
*Tony Aitchison*

12 Regional Planning in the West Midlands — 191
*John Deegan*

13 The South West – Lessons for the Future — 208
*Don Gobbett and Mike Palmer*

14 Regional Planning in South East England – A Case of 'Eyes Wide Shut' — 226
*Chris M Williams*

## PART V: CONCLUSIONS — 245

15 Some Interim Conclusions: Regional Planning Guidance and Regional Governance for the 21$^{st}$ Century — 247
*John Glasson*

# List of Tables, Figures and Boxes

| | | |
|---|---|---|
| Table 3.1 | Major British Planning Acts, and Policy Initiatives 1997–2000 | 36 |
| Table 3.2 | The Planning Framework for the UK in 2000 | 38 |
| Table 3.3 | Ten Scenarios for Spatial Planning in the UK | 40 |
| Table 3.4 | Planning Issues Requiring a National Context and Perspective | 41 |
| Table 3.5 | Alternative Institutional Models for Regional Governance and Spatial Planning | 42 |
| Table 3.6 | The Welsh Planning System in 2000: A Comprehensive Planning Framework for Wales | 45 |
| Figure 3.1 | NUTS Level 1 Areas in the European Union | 49 |
| Table 3.7 | A Typology of Regional Government in Current 15 EU Countries | 50 |
| Table 3.8 | A Typology of Regional Government in 10 Central and Eastern Europe Applicant Countries of EU as at 2000 | 51 |
| Figure 5.1 | Map of Government Offices for the Regions | 75 |
| Table 5.1 | Key Stages in Regional Planning Guidance (RPG) Preparation, Showing Role of Government Office/Secretary of State | 77 |
| Table 5.2 | North West Regional Planning Guidance: Commissioned Research Studies | 78 |
| Table 5.3 | Main Issues Raised in GONW's Response to Draft RPG for the North West Region | 83 |
| Table 7.1 | Items for RPBs to Consider Including in their Transport Strategies | 103 |
| Table 8.1 | Three Views of Regional Infrastructure | 113 |
| Table 9.1 | Proposed Annual Housing Rates for South East England | 133 |
| Table 9.2 | Changes in Water Quality and Attractiveness of Area for Developers | 139 |
| Figure 10.1 | Content and Structure of Regional Plans and Strategies | 156 |
| Figure 11.1 | Regional Strategy/Policy Framework | 169 |

| Figure 11.2 | The Challenge of Achieving Integrated Policy Development at the Regional Level | 171 |
| --- | --- | --- |
| Figure 11.3 | Key Elements and Stages in an Integrated Approach to the Development of Regional Policies and Strategies | 175 |
| Figure 12.1 | The West Midlands Region | 192 |
| Table 12.1 | New Processes of Regional Planning Guidance (RPG) | 205 |
| Figure 13.1 | Key Diagram of Draft Regional Planning Guidance for the South West | 210 |
| Figure 14.1 | Key Diagram of Draft Regional Planning Guidance for the South East | 229 |
| Figure 15.1 | Network of Principal Government Agencies at National, Regional and Local Level | 251 |
| Table 15.1 | Regional Planning Guidance Time Schedule | 253 |
| Figure 15.2 | Regional Planning Guidance/Draft Guidance: Economic Comparisons | 254 |
| Figure 15.3 | Regional Planning Guidance/Draft Guidance: Development and Land Use Comparisons | 255 |
| Figure 15.4 | Problems of Integration | 256 |
| Figure 15.5 | Layers of Policy Influence | 260 |
| Figure 15.6 | Dimensions of Policy Influence | 262 |

| Box 9.1 | Work of the Environment Agency with the Planning System | 126 |
| --- | --- | --- |
| Box 9.2 | Advice by the Environment Agency to SERPLAN on the Regional Planning Strategy | 131 |
| Box 9.3 | Deprivation Indicators in a Single Regeneration Budget Bid in the Lea Valley Area, London | 140 |

# List of Contributors

**Tony Aitchison** is Assistant Director (Strategy), East Midlands Regional Local Government Association.

**Jeremy Alden** is Professor of International Planning Studies, Department of City and Regional Planning, and Pro Vice-Chancellor, Cardiff University.

**Mark Baker** is Lecturer in Town & Country Planning, School of Planning & Landscape, University of Manchester.

**Allan Cochrane** is Professor of Public Policy and Pro Vice-Chancellor Open University.

**John Deegan** is Director of Planning, Transport & Economic Strategy, Warwickshire County Council

**John Glasson** is Pro Vice-Chancellor (Research & Consultancy) and Head of School of Planning, Oxford Brookes University.

**Don Gobbett** is County, Regional and European Strategy Manager, Dorset County Council.

**Peter Headicar** is Reader in Transport Planning in School of Planning, Oxford Brookes University.

**Hugh R Howes** is Regional Strategic Planner, Environment Agency – Thames Region.

**Greg Lloyd** is Professor of Town and Regional Planning and Head of the School of Town and Regional Planning, University of Dundee.

**Tim Marshall** is Senior Lecturer in School of Planning, Oxford Brookes University.

**Simon Marvin** is Professor of Sustainable Urban and Regional Development, Co Director of Centre for Sustainable Urban and Regional Futures, University of Salford.

**Mike Palmer** is Team Leader (Strategic Planning) with Plymouth City Council.

**Peter Roberts** is Professor of European Strategic Planning at the Geddes Centre for Planning Research, University of Dundee.

**Martin Stott** is Head of Economy and Environment, Oxfordshire County Council.

**Chris M Williams** is Chief Officer of Buckinghamshire County Council.

# Acknowledgements

We are grateful to the Economic and Social Research Council for funding the Research Seminar series on Regional Planning and Governance in England, at which many of the papers here originated (Award number R451 26 4860 98). We would like to thank all those who contributed papers to the series and who attended the seminar meetings. Thanks also to our colleague Elizabeth Wilson, who helped with seminar organisation and the editing of this book. Special thanks go to the contributors of the chapters published here, especially to those non academic 'reflective practitioners' who managed to meet (more or less!) our deadlines. We are especially grateful to Maureen Pether, who carried the burden of apparently unending alterations to the evolving text.

# List of Abbreviations

APLE Area of plan led expansion
AWM Advantage West Midlands
CBDC Cardiff Bay Development Corporation
CBI Confederation of British Industry
CEC Commission of the European Communities
CHP Combined Heat and Power
COI Central Office of Information
COSLA Convention of Scottish Local Authorities
CPRE Council for the Protection of Rural England
DBRW Development Board for Rural Wales
DCMS Department of Culture, Media and Sport
DETR Department of Environment, Transport and the Regions
DfEE Department for Education and Employment
DOE Department of the Environment
DoH Department of Health
DSS Department of Social Security
DTI Department of Trade and Industry
ESRC Economic and Social Research Council
EMDA East Midlands Development Agency
EMRLGA East Midlands Regional Local Government Association
ES Employment Service
ESDP European Spatial Development Perspective
EU European Union
FDI Foreign Direct Investment
GLA Greater London Authority
GLC Greater London Council
GO Government Office
GO East Government Office for the East of England
GOL Government Office for London
GONE Government Office for the North East
GONW Government Office for the North West
GOR Government Office for the Region
GOSE Government Office for the South East
GOSW Government Office for the South West
GOWM Government Office for the West Midlands
HBF House Builders Federation

HC Housing Corporation
HO Home Office
HQ Headquarters
IRO Integrated Regional Office
IRS Integrated Regional Strategy
LAW Land Authority for Wales
LEA Local Education Authority
LEAP Local Environment Agency Plan
LEC Local Enterprise Company
LGA Local Government Association
LPAC London Planning Advisory Committee
LTP Local Transport Plan
MAFF Ministry of Agriculture, Fisheries and Food
MPG Minerals Planning Guidance
NAW National Assembly for Wales
NDPB Non Departmental Public Body
NHS National Health Service
NHSE National Health Service Executive
NLGN New Local Government Network
NPPG National Planning Policy Guidance
NSPF National Spatial Planning Framework
NUTS Nomenclature of Units for Territorial Statistics
NWCG North West Climate Group
NWDA North West Development Agency
NWRA North West Regional Association
PAER Priority Area for Economic Regeneration
PGW Planning Guidance Wales
PIU Performance and Innovation Unit (Cabinet Office)
PPG Planning Policy Guidance
PMM Plan, Monitor and Manage
RCU Regional Co-ordination Unit
R & D Research & Development
RDA Regional Development Agency
RES Regional Economic Strategy
ROSE Rest of the South East
RPB Regional Planning Body
RPG Regional Planning Guidance
RSDF Regional Sustainable Development Framework
RTPI Royal Town Planning Institute
RTS Regional Transport Strategy
SBS Small Business Service

SCEALA Standing Conference of East Anglian Local Authorities
SD Sustainable development
SDA Scottish Development Agency
SEEDA South East England Development Agency
SEEDS South East Economic Development Strategy
SEERA South East England Regional Assembly
SERPLAN South East and London Regional Planning Conference
SoS Secretary of State
SPD Single Programming Document
SPG Strategic Planning Guidance
SRA Strategic Rail Authority
SRB Single Regeneration Budget
SSI Social Services Inspectorate
SWRDA South West of England Regional Development Agency
SWRPC South West Regional Planning Conference
TAN Technical Advice Note
TEC Training and Enterprise Council
TPP Transport Policies and Programmes
TUC Trades Union Congress
UDP Unitary Development Plan
WDA Welsh Development Agency
WMLGA West Midland Local Government Association

# PART I:
# INTRODUCTION

# PART I.
# INTRODUCTION.

# 1 Introduction

TIM MARSHALL

**Regional Planning and the 1997 Labour Government**

Most systems of governance or regulation have an element of accretion or layering. The British planning system has been accumulating strata, faults and tensions for the better part of a century. A major period of change such as the late 1940s has been the exception, with less general additions, subtractions or twists being the norm. This book is devoted to one of these apparently less fundamental changes, which gathered pace during the 1990s, and will have its full effects during the first and second decades of the new century. We will know only later how major this shift is.

The shift can be seen to have originated in the period 1988 to 1990. Nineteen eighty-eight saw the introduction of the first Planning Policy Guidance (PPG) notes (national advice to the local planning authorities), and this was followed in 1990 by a reinforcement of the role of the Development Plan within the planning system. The rearticulation of central and local control was completed in 1990 by the (re)introduction of Regional Planning Guidance (RPG) for the English regions, under the direction of central government, but following local planning authority advice.

Specifically this collection of essays focuses on the changes introduced by the Labour government of 1997 in the system of regional planning. Although the government claimed to have a programme of 'modernising planning', this turned out to have as its most significant innovation a revised process of planning at the regional level (Department of Environment, Transport and the Regions DETR, 1998; Allmendinger and Tewdwr-Jones, 2000). These changes were encapsulated in a new Planning Policy Guidance (PPG) Note, issued in draft form in February 1999 and finally in October 2000 (DETR, 1999; DETR, 2000a).

Of equal importance to these changes were the associated reforms of other regional governance processes, particularly the Regional Development Agencies (RDAs) and Regional Chambers introduced under the Regional Development Agencies Act 1998. Together with the changes specific to the town and country planning system, these reforms created a process of making RPG significantly differing from that formed in 1990, though building on that particular addition.

These process and institutional changes were accompanied by shifts in substantive policy, throughout the first three and a half years of the government. These included especially the fields of housing, transport and business. They affected all levels of the planning system. Because of the raised salience of the regions, the intersection with RPG was, however, more significant than would have been the case a few years earlier. Thus in the housing field the debate on the location of future housing for the 'extra 4.4 million households' required for the coming 20 years often settled around regional level discussions in the late 1990s. The government shift to a 'plan, monitor and manage' approach to planning for housing was integrated into the RPG process from the South East Public Examination in 1999 onwards.

In transport the same process of regionalisation of substantive planning policy making was apparent, though with some uncertainty, given the stuttering progress of the Blair government in this field. PPG 11 established the principle that Regional Transport Strategies, integrated into RPG, would have the prime responsibility for guiding transport investment, though this is far from carried into effective practice up to the present.

The Blair government has shown considerable interest in planning for business growth, particularly in the form of promoting 'clusters' of economic development. This too was incorporated as a task of RPG, in the advice given in PPG 11, with the first clear results appearing in the final RPGs issued for East Anglia and the South East in late 2000.

In the environmental field, policy also showed elements of regionalisation, with government advice pressing work in the areas of waste, renewable energy and minerals to be incorporated into RPG. Here though another policy instrument was promoted in 2000, that of Regional Sustainable Development Frameworks (DETR, 2000b). These were drawn up in each region during 2000, and their results were to feed into RPG in due course.

In broad terms then the regional initiatives of the late 1990s constituted a significant addition to the planning system as it operated in England. This stood alongside the more fundamental changes occurring in Wales, Scotland and (potentially) Northern Ireland. These changes in turn fed back on debates in England, given the sharing of ideas within the UK planning policy community, and politically.

These elements of regionalisation layered on top of the largely unchanged system of development plan-making and control, as set by the planning legislation of 1990 and 1991. This layering generated certain tensions, possibly pushing towards future reforms in those parts of the system. For the immediate future these tensions were to be worked through, rather than resolved by some generalised reform of the whole system, whatever that might imply.

This introduction will not give any fuller presentation of the changes made to the system, as the details of these changes will be found within the chapters of this book. In particular the chapters by Cochrane, Alden, Baker and Glasson give overviews of differing kinds of the 1997 to 1999 regionalisation innovations. Those unfamiliar with the British system may do well to read these chapters first.

The book is derived from a seminar series supported by the Economic and Social Research Council, under the title 'Regional Planning and Governance in England'. Many chapters here were first presented in one form or another in that series. The book does not seek to summarise directly the results of this series, which included six day-long meetings held in Oxford during 1999 and 2000.

The two remaining tasks of this introduction are to give an overview of the chapters, as a guide to the main concerns of the writers, and to explore a little more fully some underlying questions of regional planning as it is developing in England.

## An Overview of the Chapters

Readers will soon discover that the book does not have one underlying thread. The approaches of writers differ on a whole range of dimensions. The most obvious difference may be that of practitioner and academic contributors. Planning practitioners approach regional planning from the perspective of 'the coalface'. This is not the same as saying that these contributions are unaware of the 'big picture' of the underlying questions in this policy field. On the contrary, they all set their practice concerns within sophisticated understandings of the policy directions of government and other key agencies. But it is notable that these contributions (the four in Part IV, and those of Stott and Howes) are marked by and large by a sensitivity to the difficulties and stresses within practice. In a sense they operate as lightning or 'truth' conductors to place alongside the more generalised accounts in Part II. Equally though there are chapters by academic contributors which are critical of the directions of current policy (see Headicar in particular).

One other dimension of difference in contributions which may be mentioned here is the extent to which RPG is the only focus, or is part of a wider view. In most cases the interdependence of policy fields is expressed by the fact that many other dimensions of regionalisation (or non regionalisation) are examined. Only in the chapters by Baker and those on regions in Part IV is the focus mainly, but even there not exclusively, on RPG. The old message of planning's entanglement with other policy fields is clear.

*Part II*

Part II starts with a chapter by Cochrane which challenges certain matters which are often taken for granted in the literature and the practice of regional planning and regional governance. Cochrane first situates the current changes within the broader political developments of the 1990s in the UK, then goes on to examine especially the case of the South East of England. This summarises perhaps the most substantial academic research project of the 1990s on the English regions, that undertaken by colleagues at the Open University, resulting in the book by Allen *et al.* (1998). This work presents regions as 'spatialised social relations', which as such are in continuous flux and have far from definite boundaries. The rest of the chapters in this volume may appear to turn their back to this approach to understanding regions. But the difficulties of institutional design and practice which are a continuing theme throughout may on the contrary be taken as so many witnesses to this slippery kaleidoscopic fluidity and the resulting difficulties of governments and politics to get a firm hold on the 'regional imperative' (Wannop, 1995). If Cochrane's sketch is correct, we may not be surprised at the depth of the tensions within the regionalisation agenda. One reading of almost all other chapters would be that they are negotiating stresses, spatial, institutional and sectoral, generated by the variable political actualisation of the social relations emerging across the English spatial society/economy.

The other four chapters in Part II address what may broadly be called 'institutional' matters, the frameworks within which regional planning is taking place. The chapter by Alden examines the overall development of regional planning historically, places this within the context of recent spatial policy making in the European Union, and also describes current developments in Wales. His chapter may also be taken as a starting point for reading all the other chapters, as he provides a set of alternative scenarios for the future of the UK planning system(s). He makes clear that there is no certainty in the current direction of policy. However, from the current perspective, we may guess that a scenario with strong continuing elements of both national direction and local discretion, alongside the emerging regionalism, is the most likely one. That is to say, a radical rupture seems unlikely.

The variable institutional innovation in the different component territories of the UK is further examined by Lloyd in his survey of Scotland. He notes that Scotland is showing 'institutional capacity' to innovate, but in the areas of economic development and community planning, not regional/strategic planning. Change has been apparently paradoxical, with on the face of it more advance towards integrated 'regional' policy making in Wales and Northern Ireland than in Scotland. This may seem odd, as Scotland has a powerful experience in regional and national planning developed over several decades, whereas such tradition is broadly lacking in the other two territories. This paradox deserves further exploration. It may serve as a caution in drawing conclusions about the overall devolution project at these early stages. It may also be a warning about the need to see all dimensions of planning and public policy making together, in their fully political context. It is quite possible that Scottish policy makers are more deeply aware of the fundamental tensions within truly integrated government, and hence are so far unready to launch the dramatic initiatives which coherent national and regional spatial frameworks would represent. The perhaps greater depth of, and therefore larger conflicts within, Scottish democracy may be one part of this equation.

Baker provides an account of the role of the Government Office for the Region (GOR) in RPG, as revealed during 1998 to 2000, mainly in the North West but also in the south east of England. This account is informed by an awareness of the role of sub-central agents in other European states, such as that of the French Prefect. Baker finds that the GORs are if anything consolidating their dominant position in the formulation of RPG.

Material such as this will provide the base in due course for political scientists to characterise these spaces, whether in conventional terms such as elitist, corporatist, pluralist and so on, or in the more challenging theorisation of state analysis (Marsh and Stoker, 1995). No doubt the above cautions apply, that in a fast evolving system early conclusions are highly risky. However, even after two years it is quite clear that the different starting points of the English regions have generated very different abilities to innovate. Very broadly it may be observed that those regions apparently most 'coherent' in their regional arrangements have had most difficulty in finding their feet (South East, West Midlands), whilst those renowned for 'not getting their act together' have advanced rather rapidly to some kinds of integrated regional approaches (Yorkshire and Humber, East Midlands).

Stott's chapter stands apart in a sense, taking a more agnostic stance in relation to English regionalism. It takes the temperature of the arguments in the Blair government about future arrangements for local and regional governance, and finds that there is as much likelihood of a shift to a 'city regional' agenda as further moves towards regionalisation as portrayed in the rest of this book. These observations are reinforced by the portrayal of the spatially highly diverse policy making of the government up to the present, with ministerial empire building on anything but consistent regional lines as evident as any other tendency.

Here then we may sense the emerging complexity generated by the complexity of the 'spatialised social relations' of Cochrane. In particular the continuing drive to centralisation which has been a feature of all governments in England in the last 20 years may be taken as evidence of the vital relevance of some parts of the territory to any government. Writing from the 'Thames Valley', England's sunbelt, Stott is as aware as Cochrane of the way 'success' in this region is seen as key both to the country's economic future and the government's electoral fate. If RPG is ever to be genuinely handed over to regionally autonomous institutions in England, this will occur late, if ever, in the South East. The tensions with an autonomously decided strategy for London may become considerable.

## Part III

The four chapters of Part III represent only dips into the real range of what is possibly going to become 'regional spatial planning'. There are some common threads though within the three themes addressed. Each has witnessed both step changes in aspirations for policy, and great difficulties in the path of putting these aspirations into any sort of practice. Headicar describes the potentially very significant moves towards the regionalisation of transport policy, but equally shows how so far other equally innovative policy drives, particularly on Local Transport Plans, and in relation to national agencies (Highways Agency, Strategic Rail Authority) are going off at other tangents. As in other fields it is the combined effects of policy making that is key (at all levels and through all instruments), and the non-integration of government initiatives is as visible in transport as anywhere.

Marvin's chapter is exploring out on the frontier of the new themes that might be addressed by RPG, or related regional policy documents, in the future. By examining three recent regional strategies in the North West (RPG, Regional Economic Strategy and one in relation to climate change), he brings out the very different assumptions about the futures for infrastructure providers. He shows that all these strategies have significant implications for infrastructure, but that this is often buried. This may be profoundly problematic, whether from environmental, social or economic perspectives. In all sorts of ways these regions, or parts of them, will be living on their infrastructures of energy or water in very fundamental ways, over long time scales, yet regionalisation has so far hardly entered this key policy territory. The fact that in this case the 'partners' in public policy are large private corporations is also noted by Marvin, as it was by Headicar in the transport field. This is a theme deserving further study.

The chapters by Howes and Roberts both address the potential contribution of the regionalisation policies to environmental sustainability. Howes uses his privileged vantage point in the Environment Agency. The Agency is probably the key public body in the effort to achieve environmental gains, although since its formation in 1996 it has probably not been able to wield the force that some might have hoped for (House of Commons Select Committee on Environment, Transport and the Regions, 2000). Howes describes the considerable advances made in the South East regional planning processes, as well as the difficulties encountered in making more sustainable policies stick. He also provides an initial view of the early stages of policy making in London after the (re)formation of a London-wide elected authority in 2000. Though London will also be a case *'sui generis'*, many in England will be watching the making of its Spatial Development Strategy, to get a sense of how a democratically elected body, with thus enhanced legitimacy, may relate social, environmental and economic imperatives.

Roberts takes a long historical look at the scope for sustainable, or 'balanced' development through regional planning. He finds a strong track of thinking from the early twentieth century, particularly by Howard and Geddes, and sees the innovations of the 1997 Labour government as opening the potential to make significant advances towards environmentally desirable development. However, he notes several difficulties in moving effectively in this direction and concludes by laying out a path during the next decade which he feels would overcome these remaining barriers.

These sectoral studies highlight the issue always lying in the path of any serious project of regionalisation: how far will other policy making processes allow the new spatial dimension to become a central consideration in policy? At present, and in the areas examined here, the answer is, not much. National policy drives remain centrally important, and regional documents such as RPG are thoroughly integrated within these central frameworks. However, if the phrasings of PPG 11 are to become reality, we may expect some shifts in each of these sectors as in others not treated here. Otherwise any sort of horizontal integration at the regional level will remain only an unsatisfied aspiration.

*Part IV*

Chapters in Parts II and III take their main starting points in particular regions – Cochrane, Stott, Headicar and Howes in the South East, Baker and Marvin in the North West. Part IV contains accounts focussing directly on regional planning processes in the four main southern and midlands regions of England.

The East Midlands has been held up, not least in PPG 11 itself, as having created an effective form of policy integration during 1999 and 2000, in the form of the Integrated Regional Strategy (IRS). This is the focus of Aitchison's account. Probably this has been the most debated issue of the regionalisation process in England since 1997 – how can a disintegrated process be avoided, given the disjointed manner of the governmental changes? The strengths of the IRS speak for themselves in the East Midlands account. Possible weaknesses or areas for further innovation are less clear at this stage. Inevitably the proof will lie in the implementation of the particular regional strategies, above all RPG and the regional economic strategy. The conditions for success in drawing up the IRS and ensuring other strategies match it may be different from the conditions necessary for effective implementation of each strategy. These conditions are doubtless in both cases both political and 'technical'. Possibly the sharper the decisions become, that is the nearer to affecting directly particular interests, the more important the political dimensions become.

Deegan gives a clear and nuanced picture of the current challenges for RPG in the West Midlands, one of the perceived leaders in regional planning in the 1990s (Finney, 1999). Review of RPG only began early in 2000 in the West Midlands, so progress there will give clear signals as to the practicability or otherwise of RPG undertaken totally under the PPG 11 regime. Deegan does not minimise the difficulties faced under the new circumstances, though he catalogues the region's strengths as well.

Gobbett and Palmer give a lively account of the view from the front line, as the regional planning exercise of 1996 to 2001 draws to its close in the South West. They show some aspects of the 'old' (pre PPG 11) system, and then the transitional difficulties encountered in moving to the new arrangements. The South West is widely regarded as a 'difficult' area for regional planning with little commonality of view between the Bristol region, the far south west and the Bournemouth area: difficulties different from those of the South East but perhaps allowing the two to compete for the least easy ground for RPG. The difficult regions are precisely those which test the possibilities of the new system. Gobbett and Palmer show up many of the potential difficulties, including those of GOR-Regional Planning Body (RPB) relations, of participation at the Public Examination, applying 'plan, monitor and manage', the role of the Public Examination Panel, the relationship of strategic authorities and the RPB (or more generally the constitution of the RPB) – amongst others. They confess to mixed feelings as to the prospects for PPG 11 as the new base for regional planning, and make a number of suggestions which they feel would make it work better.

Williams was, like all the other authors in this section, a leading actor in the RPG for his region, in his case as lead technical officer in the London and South East Regional Planning Conference (SERPLAN) process, though outside the SERPLAN staff. His view of 'SERPLAN's last stand' is therefore extremely valuable to regional planning students. As SERPLAN was wound up in December 2000, handing over the baton of regional planning to three separate bodies working for the Greater London Authority and the South East and East of England Chambers, its last RPG will have to stand as a memorial to its work, which began nearly 40 years earlier. This may prove to be the 'last best' chance for effective spatial planning of the region, given the uncertainty over future cooperation arrangements for the three successor bodies.

It is perfectly clear that experience as chronicled here varied considerably between the English regions in the late 1990s. Almost certainly inclusion of the four other English regions would reinforce this point, although such inclusion might allow some generalisation about differences and perhaps some clustering into types: possibly an element of north and midlands versus south, and, largely coinciding, easier versus more difficult contexts. However, such sub-typing would need great care. This is partly because the variation is almost certainly complex and depends in some respects on specificities of passing conjunctures (in party politics, in interest group politics, in individual officials' capabilities and so on). Also, there are very likely important commonalities of experience between regions, and given surface variation these may be more difficult to tease out.

For the moment readers will draw their own conclusions from these revealing case studies. One observation which those familiar with the regional planning scene might agree on, is the wealth of skill and innovative capacity present particularly within the planning system, and there especially amongst officials. Such skill at operating a changing system may be only matched by the stresses and frustrations imposed by the context in which they have to work. The sensation that this public planning resource is greatly undervalued may be limited to a few within the field itself and in planning schools; if so, this is unfortunate. The implications are clear: proper resourcing, probably by creating permanent regional planning teams offering good career prospects and allowing the further development of a national pool of expertise.

## Some Central Tensions

Finally in this introduction some perhaps more fundamental issues will be broached. These are more analytical than prescriptive. Prescription is mainly reserved for the concluding chapter. These issues will be treated under the heading of tensions within the current regional planning system. Those examined here are those of scale, of sector and of politics, though these are evidently thoroughly intertwined. Another way of phrasing these tensions is that they are around the questions of why regional planning, what (kind of) regional planning, and how (to do) regional planning.

Some readers may note the relatively limited theoretical base visible here. This does not apply to all chapters, and there is of course implicit theory through all the texts. However, it is true that regional planning is in general a field which has been weakly theorised, to an extent greater even than planning as a whole. The main recent English exception to this observation is the book by Vigar *et al.* (2000), which examines, to some extent, regional level planning in the mid-1990s, using an 'institutionalist' approach. Other recent articles by Murdoch (1999) constitute another important exception, looking at strategic and regional planning processes with help from Foucaultian perspectives in particular. It is to be hoped that the relative scarcity of theory will begin to be remedied by work now underway within research programmes financed by the Economic and Social Research Council (particularly that on Devolution and Constitutional Reform). In the meantime the 'near to practice' approach evident here will remain the norm.

*Regional Planning and Scale*

One of the fundamental 'non-questions' of this field, as of many, is why regional planning needs to exist at all. There is probably little agreement even amongst planning practitioners as to the nature of a 'regional planning gap' in England. The gap might be seen, in simple terms, to rest on (1) the general argument that social, environmental and economic concerns can be addressed with added value at the English regional level, and/or (2) that policy for particular sectoral concerns (historically housing development, sometimes broader urbanisation and infrastructure proposals, in the 1990s a wider range of issues) is also usefully tackled at that level. The impetus of 1997–2000 probably contained elements of all these arguments, but with a special emphasis on regional economic development. The call of the current leader of the Conservative Party to 'return planning to local control' shows that the regionalisation process is not based on a full political consensus.

This question of 'why' is in part a question of the relevant scale of state action or more generally of public steering which is appropriate under current circumstances. An adequate answer to the question would therefore require an adequate characterisation of those circumstances, obviously as well as values, goals and so on. Here no such answer is attempted. The aim is only to pick up some of the dimensions of scale raised by contributions to this book.

One recurring theme, if often hidden, is the tension between action at national or at regional level. The new RPG is intended to be a bottom up model, if the government's press release for PPG 11 in October 2000 is to be believed ('local people to lead on regional planning'). But ministries, Government Offices for the Regions and national agencies all remain of central importance in the process. The system of PPGs created in the 1990s now guides planning practice at all levels and on most key policy issues (Tewdwr-Jones, 1997). The chapters here by Baker, Stott, Gobbett and Palmer, and Williams all bring out this tendency to retain control in central government.

Theoretically we may relate this to the tensions within the state form in the UK. Articles by Brenner (1999) and MacLeod and Goodwin (1999) are especially useful here. They argue that globalisation is one force leading to a new *power geometry*. This includes a *scalar fix* of a different form from the nation state politics of the past. In the UK the governments of the last 20 years have partially recomposed the way issues are addressed, changing the levels at which policies are formed (or not formed). This is not necessarily desired by political leaders, because within all governments and ruling elites there have been differing goals. The results, however, are none the less real. Brenner refers to *new state spaces* and *new city spaces*; regions are a bit of both, but with some characteristics of their own. Brenner (1999, p. 445) stresses regional articulation frequently, for example: 'urban economic policy is being linked ever more directly to diverse forms of spatial planning, investment and regulation on regional scales' and further on (p. 446) 'the scales of urban and regional territorial organisation are becoming ever more crucial at once as regulatory instruments of the state and as sites of socio-political conflict'.

*Jumping scales* can occur often enough with this changing dynamic. Decisions can go from the south east to the Cabinet, from Longbridge to Brussels, from a transport multi modal corridor study up to the central ministry and down to the level of a few fields. This greater fluidity may be seen as increasing the roles of government, but also constraining those roles on occasions by variably mobilised local and regional constituencies. This especially reinforces the tendency to continuous regional planning, an inability to disentangle political scale actions for as long as one, let alone five or more years.

Within this new institutional and political framework, practice by actors is still in the process of forming a more fixed system, an assumed way of doing things. Innovation in each region in regional planning, and in regional governance more broadly, is central to this, given that players hardly know the rules (draft PPGs etc) and so everyone can 'cheat' as may serve their interests – planning conferences, Local Government Association (LGA) regional bodies, GORs, Inspectors, Secretaries of State. It is likely that each region will gradually settle down to a modus operandi, influenced but not determined by other regions. Discourses will have to become shared, in order to produce operable strategies and allow action.

In all this, *strategy* is central. The emphasis on 'strategic relationalism' of Jessop (e.g. 1997), or more simply on Wenban-Smith's 'strategic opportunism' (2000) makes great sense in this field of public policy. Good strategists, whether officers, elected politicians or pressure group actors, will reap great benefits, much more so than in fields with more settled conventions and roles.

Another scale tension is that between the local and the regional. Here the planning system is unsure of itself, with its daily bread and butter consisting of development control and managing the development plan system in each locality, but the regional level now raising tempting prospects of a wider rationality. At any rate development plans still matter; a recent study of the metropolitan planning system from 1986 to the late 1990s reminded us how significant this level can be (Roberts *et al.* 1999).

Here Gobbett and Palmer touch on the stresses which the regional-local fault line causes in the South West, and Deegan notes the issue of avoiding the territory of Structure Plans and Unitary Development Plans (UDPs). Headicar brings out the potential difficulties created by a new local planning instrument, Local Transport Plans, themselves little integrated with Development Plans. Stott raises the same questions with other new local bodies, the Learning and Skills Councils, and more generally, like Headicar, points to the significance of sub-regional areas such as the Thames Valley.

If we understand the fluidity and variability of contemporary space relations (in Cochrane's terms or others), these tensions will also not come as a surprise. The variable constitution of local governance units (reinforced by the partial and chaotic unitaryisation of English authorities in the mid 1990s), and the variable regrouping of political actors (some pressure groups very much established at regional level, some interests far less likely to become an effective presence at any but the local level), serve to create a systemic unevenness between who can influence what at what level. In the absence of elected regional government and with the growing presence of quangos (RDAs above all) at regional level, this is a recipe for a regional corporatism of a byzantine form, as Vigar *et al.* (2000) discuss. In one form or another planning is used to much corporatism in its normal dealings, but its continuing commitment to working with local politicians and with local public participation may make life uneasy in this new local – regional geometry.

The schematic description above suggests some big tensions being experienced on the ground, variably in each region. These tensions call out for resolution, by some clear rule setting, or some further institutional reform (backwards or forwards). There is little sign, though, that a much more solid fix of scales is likely to emerge in the short or medium term. Extreme strategic agility may prove necessary for some time in English regional planning.

*Regional Planning and Sectors*

PPG 11 seeks to broaden the application of regional planning. Often in the early 1990s RPG was not a lot more than housing numbers expressed for counties. Practice was moving beyond this as the decade progressed, with more concern for large employment sites and for some dimensions of environmental sustainability. In principle PPG 11 specifies that a real spatial framework for the development of each region should be prepared, based firmly on a transport strategy. Waste and energy matters are also meant to be more integrated, and housing concerns are to go beyond pure numbers to some policy on the kinds of housing being built. The majority view in the seminar series was that the broadening of RPG to incorporate a much fuller range of social, environmental and economic concerns was indeed desirable.

But there were significant doubts about some aspects of this broadening, with suggestions that for example many social policies are best applied at either local or national level, and that the same can be argued for much economic and environmental policy, given English circumstances.

In this book only relatively few chapters tackle this crucial widening of the agenda – Headicar on transport, Marvin on infrastructure, Howes and Roberts on certain aspects of the environmental agenda (water, economic development). Other aspects treated more fully in the 1999–2000 seminar series were those of health and social exclusion, and how voluntary organisations were organising themselves to participate in regional level debates.

What appears to be emerging is considerable divergence in the extension to new sectors genuinely incorporated into RPG. This is definitely a 'multi-speed' process. Housing and regional economy are generally more fully on board, but the struggle to present a more integrated spatial strategy, with transport, is only just beginning to have effects. How far environmental concerns will become central, beyond the spatial strategy/transport key area, remains to be seen, whether shepherded in by an Integrated Regional Strategy or by the Regional Sustainable Development Frameworks.

Regional Scoping Reports on climate change are now complete for many regions, but their integration into regional strategies remains a major challenge. There is still uncertainty in the planning community about what the incorporation of issues like health, education and culture would really mean at regional level. One suspects that any move to make real decisions through RPG in these areas could create tensions with the respective ministries and their associated policy communities. In any case, as Gobbett and Palmer argue here, it may be that RPG should not seek to cover the whole range of topics recommended by PPG 11, in every region.

To a large degree this quite uneven colonisation of new territory by RPG takes us back to the very skewed regionalisation of the Labour programme, with the economy prioritised above all, institutionally and financially. This in turn points again to the way in which the state form and more specific governance arrangements (partnership etc) are being reformed, with a complex dynamic of scale and sector.

*Regional Planning and Politics*

The state form is changing in the UK, and this is feeding directly into the dilemma-concentrated task of spatial planning. There is all kind of crisis displacement and reworking of power geometries going on. Whilst it may seem that the relatively humble role of RPG (to set housing allocations) did not need such a grand framing, the moves to broaden to much more ambitious spatial strategising, on a more inclusive political foundation, raise quite different issues. For the first time the regional space may become a political space in the more open sense. Whilst central government may in due course be able to close off this drift, the scope for some regional proactiveness beyond the housing trap does exist currently. The current 'state hegemonic project', in Jessop's terms, evidently includes a certain movement in England's power geometry, and for the moment regional planners and other regional actors cannot escape this fact.

The power geometry built into the regional institutional reforms may be intended to further the formation of hegemonic regional blocs pressing for strategies of business-led competitiveness. However, other mechanisms of responsiveness embedded within some regions can easily disrupt the emergence of such blocs, and leave only a frustrating semi-vacuum in most regions (frustrating for some planners and some interests, probably not too bad an outcome for many centrally located politicians and interests).

It may by this stage be almost superfluous to refer to the political dimensions of regional planning. Both the above sections refer implicitly to the profoundly political nature of the regionalisation process in England. In a sense this has been rather invisible, unlike the open struggles in Wales, Scotland and London. In part this is a true reflection of the inability of English regional politics to 'bark' and so become newsworthy. Even the broadsheet newspapers have reported very sparsely the goings on in the English regions in the last three years, whatever the fascination of some planners and economic regeneration specialists. The main exception has been at certain stages of the South East process, especially at the Public Examination at Canary Wharf in May 1999 – no doubt showing the southeast-centric orientation of the national press. Contributions in this book go only some way to change this perspective, with only the chapters by Stott and Williams drawing aside the political curtain a little. This reflects a long tradition in planning, born partly of the necessities of practice, that it is not polite to discuss politics openly, whether of party or pressure group form.

The peculiar regionalisation underway may press against the maintainance of this polite fiction. With elected regional government, business could continue as usual, with the assumption that key decisions were being taken legitimately where they belonged. In the absence of such formal institutionalisation, the real politics of the new regional spaces becomes central, however undiscussed. It is clear that some of the differences between regional practice relate to the nature of party politics in each region, with different kinds of Labour control for example in different northern and midlands regions, and a particular brand of Liberal Democrat influence still just dominant in the South West. This has affected the decisions on whether RPG should be controlled by the regional branch of the Local Government Association, or by the newly formed Regional Chamber, and the nature of relations with the RDA.

Equally centrally, interests have organised to different effect in each region. Probably the main actors now, regional business coalitions represented mainly through the Confederation of British Industry (CBI) and Chambers of Commerce, have different organisation and impact in each region, and so relate to their 'partner' RDA in different ways. The same applies to other key actors such as the Council of the Protection of Rural England (CPRE) or Friends of the Earth.

It also matters that the various outposts of the central government, through the GORs and the national quangos (Environment Agency, Countryside

Agency and so on) have somewhat differing approaches in each region. Some GORs are more involved than others in RPG detail, at earlier stages. A change of Regional Director can also make a real difference in the approach to planning of the GOR. Again this is brought out little by chapters here. It is not easy to move beyond anecdotal evidence to back up such assertions, though research currently underway will hopefully make this clearer.

The way in which such emerging political arenas are to be analysed has not yet gelled. One favoured route is by drawing on network analysis, given the dominance of 'informal' politics and the apparent importance of the building up of relatively tight sets of personal linkages (e.g. John and Cole, 2000) for example. A key issue here is how continuous the regional planning process becomes, with the references in PPG 11 to partial revision at relatively frequent intervals. This may lead to a differently shaped political process, again possibly tending to a somewhat closed corporatism, whereby a few key actors effectively take the lead in keeping RPG up to date. On the other hand the kind of open process being attempted in the West Midlands may be successful in setting an approach which becomes ever more inclusive, through the development of regionally effective public arenas. Whatever is the case, it will surely be valuable to open up the study of the political dimensions of the regional planning process, given the distance it has moved from the (apparently) technical exercises of earlier years.

## References

Allen, J., Massey, D. and Cochrane, A. (1998), *Rethinking the Region*, Routledge, London.

Allmendinger, P. and Tewdwr-Jones, M. (2000), 'New Labour, new planning? The trajectory of planning in Blair's Britain', *Urban Studies*, vol. 37 (8), pp. 1379–402.

Brenner, N. (1999), 'Globalisation as reterritorialisation: the re-scaling of urban governance in the European Union', *Urban Studies*, vol. 36 (3), pp. 431–51.

Department of the Environment, Transport and the Regions (1998), *Modernising planning: a statement by the Minister for Planning, Regeneration and the Regions*, DETR, London.

Department of the Environment, Transport and the Regions (1999), *Planning Policy Guidance Note 11. Regional Planning*, public consultation draft, DETR, London.

Department of the Environment, Transport and the Regions (2000a), *Planning Policy Guidance Note 11. Regional Planning*, DETR, London.

Department of the Environment, Transport and the Regions (2000b), *Guidance on Preparing Sustainable Development Frameworks*, DETR, London.

Finney, J. (1999), 'Regional planning to the fore', in D. Chapman (ed.) *Region and Renaissance: Reflections on planning and development in the West Midlands 1950–2000*, Brewin Books, Studley, pp. 185–92.

House of Commons Select Committee on Environment, Transport and the Regions (2000), *The Environment Agency: sixth report*, Stationery Office, London.

Jessop, B. (1997), 'A neo-Gramscian approach to the regulation of urban regimes: accumulation strategies, hegemonic projects, and governance', in M. Lauria (ed.) *Reconstructing Urban Regime Theory*, Sage, London, pp. 51–72.

John, P., and Cole, A. (2000), 'When do institutions, policy sectors and cities matter? Comparing networks of policy-makers in Britain and France', *Comparative Political Studies*, vol. 33 (2), pp. 248–68.

MacLeod, G. and Goodwin, M. (1999), 'Space, scale and state strategy: rethinking urban and regional governance', *Progress in Human Geography*, vol. 23 (4), pp. 503–27.

Marsh, D. and Stoker, G. (1995), *Theory and Methods in Political Science*, Macmillan, London.

Murdoch, J. (1999), 'Governmental rationalities in conflict: a reflection on the changing relationship between space and time in planning for housing', paper presented to Planning Research 2000 conference, LSE, March 2000.

Roberts, P., Thomas, K. and Williams, G. (1999), *Metropolitan Planning in Britain. A Comparative Study*, Jessica Kingsley, London.

Tewdwr-Jones, M. (1997), 'Plans, policies and intergovernmental relations: assessing the role of national planning guidance in England and Wales', *Urban Studies*, vol. 34, pp. 141–62.

Vigar, G., Healey, P., Hull, A. and Davoudi, S. (2000), *Planning, Governance and Spatial Strategy in Britain*, Macmillan, London.

Wannop, U. (1995), *The Regional Imperative*, Jessica Kingsley, London.

Wenban-Smith, A. (2000), 'Towards strategic opportunism', *Planning*, 31 March 2000.

# PART II:
# REGIONAL REALITIES
# AND
# REGIONAL INSTITUTIONS

# 2 Thinking about the English Regions

ALLAN COCHRANE

The official rediscovery and, indeed, celebration of a regional dimension to British – and particularly English – policy and politics is a relatively recent phenomenon. In the 1980s regional planning passed through a 'dark age' following the abolition of the economic planning councils, although there was a process of understated revival through the early 1990s (Simmons, 1999, pp. 160–3). There was fundamental government scepticism about the value of planning – particularly regional planning – which was seen as an unnecessary bureaucratic constraint on the vitality of market processes. Although the Scottish, Welsh and Northern Ireland Offices survived even through the lean years, the first (rather weak) shoots of officially sanctioned English regionalism were only seen in the creation of the Government Offices in the mid-1990s. Even they were initially met with suspicion by those committed to local and regional autonomy – perceived merely as regionally based branches of central government, deconcentrated, rather than decentralised government (Mawson and Spencer, 1997). Today the existence of a regional tier with the potential to shape and develop policy is increasingly simply taken for granted.

The so-called nation regions of Wales and Scotland now have devolved governments and Northern Ireland, too, has its own elected Assembly. In England, the Department of the Environment has been relabelled the Department of the Environment, Transport and the Regions (DETR) and there is a Minister for the Regions. Some state activities are now the responsibility of Regional Development Agencies (RDAs), which are required to develop regional strategies focused on economic development, skills development and regeneration. Some significant state funds (including the Single Regeneration Budget) are to be channelled through them. It looks as if we may even be on the track towards elected regional government of one sort or another, and the European Union remains committed to the pursuit of a Europe of the Regions. The chapters of this book explore some of the implications of the re-emergence of regional planning, considering governance and institutional issues, as well as the nature of regional planning in practice and the regional dimensions of planning for sustainability, transport planning, social inclusion and the utilities.

## Looking for Regions

What does this rediscovery of regions amount to? It is easy enough to talk about regions, but what are they?

For some people, even asking this question may simply seem irrelevant. The new frameworks are there and need to be worked with. In any case, it is often taken for granted that regions exist and, indeed, that in England they pretty much accord with the standard regions already associated with the government offices and assumed by the DETR. The boundaries of England's planning regions have, after all, been in place through most of the last century and have been used as a basis for generating comparative data for many years. As long ago as the 1930s regional divisions were used as the basis for developing policy, which focused on the Assisted Areas. They continue to underpin every discussion of the North-South divide, even as other statisticians seek to subvert them by providing micro-analysis that shows the divisions that exist within them (see, for example, SEEDS, 1987 and Breugel, 1992).

Discussions about regions, therefore, generally start from their official representations, the ways in which they have been defined for the purposes of the government statisticians, and the organisation of government offices. They are presented as fixed entities, made up of an aggregation of counties, even if there are sometimes debates about the detailed drawing of boundaries between them. The main emphasis is on how to bring economic or political life to these statistical regional constructs.

From a planning point of view, the boundaries are used as a major starting point, although there are sometimes attempts to build sub-regional plans that cut across them. The new RDAs take them for granted, and their activities help to construct them as identifiable entities (see, for example Jones and MacLeod, 1999, for an analysis that locates them within a process of the re-scaling of the state). It is widely recognised that most of the English regions around which professional activity is organised (whether in economic development, planning or relating to other government functions) have little clear meaning for most of those who live in them. Although this may vary between regions, individuals rarely identify themselves with the regions to which they are assigned in the same way as they do with their nations, counties, cities or neighbourhoods of residence (or birth). While there seems in recent years to have been a significant (albeit regionally differentiated) shift in the proportion of people resident in England who identify themselves as English rather than British, there has been no similar increase in regional identification (Curtice and Heath, 2000).

None of this may matter much, of course, if all that is happening is that the institutions of government are looking for some more or less convenient way of dividing responsibilities between them, as a response to the fear that excessive centralisation may lead to inefficiency, as well as a lack of democratic and other forms of accountability. In other words if all that is happening is that different levels of government are being assigned to different territorial scales, then it does not matter too much how those territories are defined (see, for example Rhodes, 1988).

If, however, claims are being made for the coherence and identity of the territories, then obviously it matters rather more. So, for example, the logic of deconcentration through Government Offices might be seen as one that fosters working between government departments at an appropriate level, while remaining an organisational arrangement within central government. The logic of RDAs, however, seems to be a rather different one, which draws its justification from the notion of the region as a more or less coherent economic and social space whose internal linkages make it possible to plan its economic development as a whole. That seems to be the direction currently being taken in the official discourse of government as well as the practice of those working at regional level. In that context, the questions do matter.

**New Regional Geographies**

The rise of economic globalisation has called into question the central position of national economies (perhaps most extremely in the work of Ohmae, 1990, who identifies the emergence of a 'borderless world', and Castells, 1996, who seeks to chart the rise of a 'network society'). That does not mean, though, that there are no territorially based economic divisions of labour. Indeed, Ohmae points to the rise of city regions as the emergent alternative to nation states, while Castells similarly charts the rise of what he describes as 'global megacities'. Both the city regions and the megacities are said to cut across national boundaries in their economic and cultural influence. Globalisation, it is argued, opens up significant new possibilities for the regions that were previously trapped within the prisons of their nation states. A much higher degree of regional economic autonomy is promised.

This makes it all the more important not simply to take the administrative divisions – the lines on the map – for granted. One way of moving beyond the lines is to consider regions as what Allen *et al.* (1998) refer to as spatialised social relations, that is as places which are actively constructed by the economic and social networks and linkages within them and which connect them to other places. In most political debates on regions, in practice emphasis is often placed on their constructions as economic activity spaces, but – as Amin and Thrift (1994) stress in their discussion of the significance of 'institutional thickness' – this may include the cultural and institutional relationships that underpin economic processes. Unlike the fixed entities frozen by the cartographers of the official regionalism, this approach suggests rather more fluid possibilities and implies the possibility of overlapping regions of different sizes. These are regions that may vary across time and are made and remade by human interaction (within which – of course – the cartographers, government officials and regional politicians play their own part). These regions do not have clear and permanent boundaries, and are defined by their positions within networks, which stretch out much wider nationally and internationally through a series of interconnections.

Geography – or the geographical imagination – opens up ways of understanding regions in different ways, namely, as sets of living relationships. Places are not fixed entities, handed down from above and then fixed in amber. On the contrary, they are constantly being defined and redefined through the working out of overlapping networks of social relations. From this perspective, it is essential to start by thinking in terms of relationships, of the processes by which regions are constructed and reconstructed.

The rebirth of regionalism as a political idea has been underpinned by the rise of a new regional economic geography, which suggests that regions and regional networking can help to generate economic innovation, competitiveness and prosperity. The regional scale is said to be the one at which productive interaction between enterprises, social institutions and the state can best come together to generate successful innovation and hence economic well-being – self-sustaining growth. The examples of Baden-Württemberg and Emilia-Romagna have frequently been used as examples (see, for example Cooke and Morgan, 1994; Morgan, 1994), but Maskell *et al.* (1998) also explore some of the ways in which regional specialisation may help regions to prosper in the context of economic globalisation (in what they call 'small open economies') with the help of evidence from the Nordic countries.

Storper (1997) has also emphasised the importance of the relational assets that may be embedded in particular regions, highlighting the importance of identifying the ways in which the relations that underpin some regional economies may encourage a process of shared learning between firms. Similarly in their discussion of 'institutional thickness' as a prerequisite for some forms of growth, Amin and Thrift (1994, p. 14) highlight the potential for the development of a 'mutual awareness that' the institutions 'are involved in a common enterprise'. Morgan (1997) stresses the possibility and importance of trust-building and regular interaction between key players at regional level (through a range of what he calls 'institutional routines and social conventions', p. 493) as a fundamental basis on which an innovative or learning region may be constructed. Historically regional policy in the UK has been a form of social welfare policy – encouraging a redistribution of resources from wealthier to poorer regions. Now, says Morgan, it is important to have a different starting point, in which a regional focus may make it possible to generate prosperity and growth through building institutional capacity.

This is consistent with the government's regional agenda, which places a fundamental emphasis on economic competitiveness – to 'promote sustainable economic development' (Department of the Environment, Transport and the Regions, 1997), with a 'greater focus on wealth creation and jobs' (John Prescott, quoted in Jones and MacLeod, 1999, p. 301). Fostering the competitiveness of regions is presented as a means of providing them with a secure economic base, capable of underpinning the financial well-being of their residents. Collectively the competitiveness of individual regions is, of course, also believed to provide a means of enhancing the overall competitiveness of the country. But at the same time the regional agenda is still effectively presented as a means of achieving political and economic redistribution. In the coded language of new Labour politics, a focus on the 'regions' provides a means of acknowledging the existence of economic inequality, as well as promising a programme of renewal through a form of economic self-help.

It is sometimes difficult to escape the echoes of the economic planning councils of the 1970s and even the debates about the 'distressed areas' of the 1930s, but this time the stress is clearly on regional initiative rather than a national planning regime. If only everywhere can become more competitive, then everywhere will be better off. This is noted by Tewdwr-Jones and Phelps (2000, p. 438) in their discussion of strategies to attract foreign direct investment: 'By enhancing the resources and strategic capabilities of the English regions, the government has attempted to level the uneven playing field between England, Scotland and Wales.'

The policy context is different, however, because this time everywhere is a 'region' – everywhere is apparently driven by similar needs – from the South East of England to the North of England. *Every* region is being enjoined to improve its economic competitiveness. The South East of England Development Agency (SEEDA) makes no bones about its purpose, with a strap-line that promises that it is 'Working for England's World Class Region'. Regional policy is no longer a policy for a set of 'regions', implicitly defined as those places with economic and social problems which lie outside the golden heartlands of London and the Home Counties. Instead, a single template, implicitly borrowed from the European – and specifically the German – model, albeit without the creation of elected regional governments, is to be applied across England. All RDAs are challenged to perform similar tasks, even if the resources allocated to each of them are expected to vary according to the scale of the challenges they face.

Lovering (1999) has launched a powerful critique of this approach, and the academic analysis that underlies it, which he calls the 'new regionalism'. He draws specifically on the Welsh experience, but his arguments are of wider relevance. He is sharply critical of the way in which he believes academics have taken their lead from the policy agenda of the regional economic boosterists, whose ambitions are to re-imagine their regions through the prism of the 'new' and the 'innovative' rather than seeking to deal with the stubborn realities of the old (see also Lovering's analysis of urban managerialism, Lovering, 1995). 'The policy tail', he claims, 'is wagging the analytical dog' (Lovering, 1999, p. 390). One aspect of his critique questions the widespread focus on innovation clusters and high technology as key elements in the development of policy, which may fit into an approach that emphasises competitiveness, but fails to recognise that success in these terms may actually lead to increased regional unemployment.

More fundamentally, and paradoxically perhaps, he also questions the theoretical arguments because of their distance from existing regions. He suggests that because the theorisations are used as a justification for a particular policy direction, they are often underspecified. So, for example, the notion of a 'learning region' may find an expression in theory at a certain level of generality, but it is less clear what it might consist of in practice, even if it can be used after the event to label some already successful region. It is easier to describe an existing 'learning region' than it is to show how a region might become one.

Whilst they acknowledge the value of some aspects of the 'new regionalism' (particularly as developed in the work of Storper and of Amin and Thrift), Jones and MacLeod (1999) draw similar conclusions about England's new regional structures. They argue that 'these attempts to contrive a new territorial map of institutional compromise' from above is likely to 'fail effectively to coalesce and represent the diverse set of interests that prevail within Britain's regional spaces', because 'RDAs may only serve to *rescale* rather than *resolve* Britain's long-standing regional problems' (Jones and MacLeod, 1999, p. 297). In other words, instead of 'regional' problems being problems to be resolved through the development of national policies, they are redefined as problems to be resolved at regional level, through increasing regional competitiveness.

At a policy level, Lovering's message is pretty clear. The emphasis on competitiveness as a focus for regional economic development policy is itself a problem, because it implies that success will be measured through the ability of a region to attract (or potentially grow) large firms and sustain a high technology sector (or whatever is defined as a growth sector at the time). Despite a rhetorical stress on social inclusion, it is likely, he argues, that a concern for unemployment will take second place. Instead of the currently emergent model, stress is placed on the need for forms of planning which can cut across regions and localities, in a sense revisiting the social welfare approach dismissed by Morgan (1997) (see also Allen *et al.* 1998, pp. 126–36 and Hudson *et al.* 1997, who highlight the importance of the national context in influencing regional success).

**Re-thinking the Region**

These issues are explored by Allen *et al.* (1998) with the help of a consideration of the recent history of the South East of England (based on a series of linked projects funded by the Economic and Social Research Council). Through the 1980s and into the 1990s the South East played a key role in the symbolism of Thatcherism, as an economic as well as a political project. Its apparent success – as a growth region – was explicitly mobilised as a means of highlighting the failure of the older industrial regions.

Of all the British regions, the South East is probably the one whose conceptualisation is most awkward. It surrounds London, stretching from Hastings in the South to Milton Keynes in the North. The government office is in Guildford, while London is the absent present that in many ways helps to define the region's dynamic.

There are huge differences within the region. Yet it is often discussed as a statistical aggregate, as if it had an identity created by lines on maps, within which the statistical data are collected. Since the current politics of regionalism require regional actors to emphasise the problems they face, however, there has recently been a remarkable turnaround in official representations. Where it was once left to agencies such as the South East

Economic Development Strategy (SEEDS) to point to divisions within the South East, now mainstream consultants are commissioned to identify those divisions, so that the new regional agencies have something with which to bargain in the search for state funding. In the Regional Review on its web-site, the Government Office for the South East explicitly refers to research undertaken by Llewellyn-Davies, which highlights some of the major divisions within the region as part of this process of jockeying for regional advantage (Government Office for the South East, 2000) and similar points are made in SEEDA's Regional Strategy.

The key questions tackled by Allen *et al.* (1998) are: What is or where is the South East? But also *when* was the South East? In other words, it is important not to assume that 'it' is the same or has always been the same.

They set out to identify key aspects of the ways in which the South East was defined and understood in the 1980s and 1990s and concluded that the South East was defined and defined itself as a 'growth' region and specifically a region of neo-liberal growth. Its national dominance was expressed through this definition which reflected a particular confluence of political, cultural and economic dynamics. Although it was presented as a model of deregulated growth, in practice, it relied on a high degree of state intervention both to achieve the particular forms of 'deregulation' that were driven through and tended to advantage the South East, as well as significant investment in large-scale public infrastructure, for example, associated with road construction, from by-passes to the M25. The growth of the South East was predicated on decline elsewhere in the UK, in the wake of large scale (state-sponsored) industrial restructuring. In other words, it was not possible to view the region as self-contained and simply building its own competitiveness. Its positioning within a wider national (and European) system helped to shape its growth.

Viewing the South East as a 'growth' region confirms the limitations of more static definitions, and it also highlights some of the difficulties of utilising generally available data in pursuing dynamic processes of change. There is no single 'correct' definition of the growth region. It stretches far beyond the 'standard' region for some forms of economic and cultural relationships (for example to Cambridge, through the threads connecting high technology industry, and pulled to Wiltshire by the tentacles of the luxury – 'country' – housing market). Meanwhile, within the 'standard' region there are substantial spatio-social discontinuities – holes and hot spots (the holes represented by the high water mark of Fordism in places like Luton and the Medway Towns and the hot spots represented by the M4 corridor, the airports of Gatwick and Heathrow and the new town of Milton Keynes). Indeed the region may be defined in terms of the divisions that cut across it. The region is defined by the complexity of these interconnections and oppositions, in ways that are fluid rather than final.

In the context of the South East these understandings make it easier to understand the fragility of growth, the extent to which the growth of the South East in the 1980s was self-defeating, because

- restructuring took place according to a neo-liberal agenda, and effectively reinforced the separation of the South East from the rest of the UK;
- levels of inequality within the region increased;
- new symbolic social geographies emerged which reflected and encouraged social restructuring (around 'race', class and gender) (the ROSE – Rest of the South East – was increasingly a white one);
- areas that needed growth did not get it, while those that did not need growth did get it; and
- as a result costs were driven up, particularly in the context of labour market and housing market rigidity.

As Amin and Thrift (1994, p. 17) acknowledge, the experience of the South East fits uneasily with models that point towards the importance of 'institutional thickness' as underpinning economic growth or suggest that it is only well integrated 'learning regions' that can be competitive. On the contrary the growth of the South East in the 1980s relied on the entry of new – 'untrusted' – players into previously protected market places. Many of the 'hot spots' (like Milton Keynes, Bracknell and Swindon) were weakly developed in institutional terms, while in others, such as Cambridge, the relationship between the existing institutions and the new were always much more attenuated than the 'hype' suggested. And – as argued above – it is important to remember that the 'success' of the South East was part of a wider national economic and political project.

Nevertheless, a fundamental aspect of the South-East's failure to sustain growth, either within the region or as a driver for the UK as a whole, was to be found in the limitations of its institutional base. It lacked a developed set of interlocking networks of trust, which would have made it possible to move beyond sectional interests. There was no effective means of distributing growth within the region, so there were concentrations of employment in particular areas, which pushed up the costs of transport, housing and labour. Instead of sustainable growth, there was continued development on greenfield sites on the edges of the 'hot spots'. Some places were overdeveloped, while others were left to rot. As Peck and Tickell (1992, p. 359) note: 'The region enjoyed strong economic growth as a result of its privileged place in *national* regulation strategies and *international* accumulation strategies, but subsequently was shown to lack appropriate *regional* regulatory mechanisms for the sustenance of growth.'

**Conclusion**

However enthusiastically one views the prospects of the Government Office for the South East or SEEDA it is hard to see them adding a great deal to the economic governance of a region like the South East, whose position within the UK's regional system gives it a special status, both as potential driver of and potential brake on national economic growth. Nevertheless, the experience of the South East in the 1980s certainly highlights the urgent need for regional planning of the sort discussed in the later chapters of this book. The lack of any major regional institutional or governance framework (despite the best efforts of SERPLAN) underlined the need to have some overview capable of managing growth and its implications. The lack of any regional strategy for transport, for social housing or sustainable growth acted as a powerful constraint on growth. However, not everything can be resolved at the regional level, unless regional plans are part of some wider approach or set of policies. The danger is that the new regional agenda will make it more difficult to plan across regional boundaries, at a time when the voracious demands of the South East may only be satisfied as part of a co-ordinated programme that allows and encourages a distribution of growth beyond its boundaries.

The purpose of this brief summary of research is, however, not to reach some final conclusion about the experience of the South East in the 1980s and 1990s, or to point towards answers for the twenty-first century. The conclusions drawn by Allen *et al.* (1998) may be mistaken or misguided. The important point here is not to set out our understanding of how the South East has developed, but rather to highlight the importance of thinking in terms of relationships, and of being spatially sensitive in doing so. There can be no social relation that does not have a spatial form. Regions need to be understood as spatialised social relations, and it is that which makes the use of the geographical imagination so vital in developing an understanding that brings them to life, outside the pages of planning texts and government reports. As the actions of politicians and civil servants seek to freeze regions into the boxes that have been handed down from above, it will be important to remember that the emergent activity spaces are likely to show little respect for the boundaries defined by them.

# References

Allen, J., Massey, D., and Cochrane, A., with Charlesworth, J., Court, G., Henry, N. and Sarre, P. (1998), *Rethinking the Region*, Routledge, London.

Amin, A. and Thrift, N. (eds.) (1994), *Globalisation, Institutions and Regional Development in Europe*, Oxford University Press, Oxford.

Breugel, I. (ed.) (1992), *The Rest of the South-East: a Region in the Making?*, South East Economic Development Strategy, Basildon.

Castells, M. (1996), *The Rise of the Network Society*, Volume 1 of *The Information Age. Economy, Society and Culture*, Blackwell, Oxford.

Cooke, P. and Morgan, K. (1994), 'Growth regions under duress: renewal strategies in Baden Württemberg and Emilia-Romagna', in Amin and Thrift (eds.).

Curtice, J. and Heath, A. (2000), 'Is the English lion about to roar? National identity after devolution', in R. Jowell, J. Curtice, A. Park, K. Thomson, L. Jarvis, C. Bromley, and N. Stratford (eds.) *British Social Attitudes. Focusing on Diversity. The 17$^{th}$ Report*, Sage, London.

Department of the Environment, Transport and the Regions (1997), *Building Partnerships for Prosperity*, HMSO, London

Government Office for the South East (2000), *Regional Overview*, [http://www.go-se.gov.uk/overview_t/index.html].

Hudson, R., Dunford, M., Hamilton, D. and Kotter, R. (1997), 'Developing regional strategies for economic success: lessons from Europe's economically successful regions', *European Urban and Regional Studies*, vol. 4, pp. 365–73.

Jones, M. and MacLeod, G. (1999), 'Towards a regional renaissance? Reconfiguring and rescaling England's economic governance', *Transactions of the Institute of British Geographers*, NS vol. 24, pp. 295–313.

Lovering, J. (1995), 'Creating discourses rather than jobs: the crisis in the cities and the transition fantasies of intellectuals and policy-makers', in P. Healey, S. Cameron, S. Davoudi, and M. Madanipour (eds.) *Managing Cities: the New Urban Context*, Wiley, Chichester.

Lovering, J. (1999), 'Theory led by policy: the inadequacies of the 'new regionalism' (illustrated from the case of Wales)', *International Journal of Urban and Regional Research*, vol. 23, pp. 379–95.

Maskell, P., Eskilinen, H., Hannibalsson, I., Malmberg, A. and Vatne, E. (1998), *Competitiveness, Localised Learning and Regional Development. Specialisation and Prosperity in Small Open Economies*, Routledge, London.

Mawson, J. and Spencer, K. (1997), 'The government offices for the English regions: towards regional governance?', *Policy and Politics*, vol. 25, pp. 71–84.

Morgan, K. (1994), 'Innovating by networking: new models of corporate and regional development', in M. Dunford and G. Kafkalis (eds.), *Cities and Regions in the New Europe: the Global-Local Interplay and Spatial Development Strategies*, Belhaven, London.

Morgan, K. (1997), 'The learning region: institutions, innovation and regional renewal', *Regional Studies*, vol. 3, pp. 491–503.

Ohmae, K. (1990), *The Borderless World. Power and Strategy in the Interlinked Economy*, Collins, London.

Peck, J. and Tickell, A. (1992), 'Local modes of social regulation? Regulation theory, Thatcherism and uneven development', *Geoforum*, vol. 23 (3), pp. 347–63.

Rhodes, R.A.W. (1988), *Beyond Westminster and Whitehall: Sub Central Governments of Britain*, Unwin Hyman, London.
SEEDS (1987) *South-South Divide*, South East Economic Development Strategy, Stevenage.
Simmons, M. (1999), 'The revival of regional planning. Government intentions; local reality?', *Town Planning Review*, vol. 70 (2), pp. 159–72.
Storper, M. (1997), *The Regional World. Territorial Development in a Global Economy*, Guilford, New York.
Tewdwr-Jones, M. and Phelps, N. (2000), 'Levelling the uneven playing field: inward investment, interregional rivalry and the planning system', *Regional Studies*, vol. 34 (5), pp. 429–40.

# 3 Scenarios for the Future of Regional Planning Within UK/EU Spatial Planning

JEREMY ALDEN

**Regional Governance and Planning, Past and Present**

Interest in spatial planning within both the United Kingdom (UK) and European Union (EU) has never been greater, particularly interest in the institutional and political dimensions of planning at the regional level. This chapter looks at recent changes in the spatial planning system in the UK and EU, and explores some scenarios for the future. The Labour Government of 1997 has been associated with several major changes, particularly (a) constitutional reform in devolving decision making and governance to the regional level, and (b) strengthening the European context for planning in the UK. However, after some significant initiatives, there is now some uncertainty as to the future progress of regional planning, both in terms of substantive content and pace of change.

This chapter also gives attention to changes in Wales. Whilst Wales continues to share planning legislation with England, it produces its own Planning Policy Guidance (PPG). Its PPG Wales 1999 document is currently being revised by the National Assembly for Wales' Land Use Planning Forum (of which the author is a member). There are also proposals to prepare a spatial planning perspective for Wales, equivalent to RPG in the English regions. In addition Wales is an Objective 1 area (covering 15 of the 22 unitary local authorities) in the EU Structural Funds programme for 2000–2006. The European context has become particularly important for Wales.

One of the features of planning in Britain has been its combination of a strong base in local government, and its centralised nature. Central government prepares national guidance (PPGs and RPGs), which are to be taken into account by local planning authorities in preparing development plans and in decisions on individual planning applications. Another feature of the British planning system is its discretionary nature, unlike most other planning systems in the world which follow the regulatory (i.e. zoning) model. Table 3.1 lists major Planning Acts and some post-1997 policy initiatives.

**Table 3.1: Major British Planning Acts, and Policy Initiatives 1997–2000**

| Year | |
|---|---|
| 1909 | Housing, Town Planning Act |
| 1909 | Planning Inspectorate and Appeals Procedure |
| 1919 | Housing, Town Planning Act |
| 1932 | Housing, Town Planning Act |
| 1941 | Uthwatt Report (Expert Committee on Compensation and Betterment) |
| 1947 | Town and Country Planning Act |
| 1953 | Town and Country Planning Act |
| 1967 | Land Commission Act |
| 1971 | Town and Country Planning Act and Land Commission Dissolution Act |
| 1975 | Community Land Act |
| 1976 | Development Land Tax Act |
| 1990 | Town and Country Planning Act |
| 1991 | Planning and Compensation Act |
| 1997 | Department of the Environment Circular: Planning Obligations |
| 1997 | Planning Policy Guidance Note 1: General Policy and Principles |
| 1998 | New Labour Government's Proposals for Modernising Planning |
| 1999 | Planning Policy Guidance Note 11: Regional Planning (Public Consultation Draft). Also PPG 12 on Development Plans |
| 1999 | Progress Report (DETR) 'Modernising Planning' |
| 1999 | Devolution for Scotland, Wales and Northern Ireland, and RDAs for English Regions. |
| 1999 | Publication of final version of European Spatial Development Perspective, looking at spatial balance in EU. |
| 1999 | Pre-Budget Report: Planning System required to promote competition. RPG to address economic competitiveness. |
| 2000 | Regional Chambers for English Regions. London elected mayor and planning authority. |

*Source:* Compiled by the author

'Regional Planning is an idea whose time has come' was the message of a presentation by the head of the new East of England RDA at a Royal Town Planning Institute (RTPI) conference in June 1999 (Samuel, 1999). Whilst there has been a flurry of activity in the preparation of regional strategies by both Regional Development Agencies (RDAs) and RPGs in England

(and elsewhere) during the past year, it should not be forgotten that regionalism is not a new idea. It continues interest beginning in the first decade of the twentieth century. In the later 1990s regionalism has been linked to 'regions and Europe'. However, in the first decade of the twentieth century regionalism was linked to 'regions and local government', as illustrated by the Fabian Society report advocating the reform of local government on a regional basis (Sanders, 1905). The work of the Fabian Society was followed by the publication by Patrick Geddes (1915) of *Cities in Evolution*, emphasising the need for a regional approach to local government to meet the needs of a city-region style of life. In 1919, C.B. Fawcett published *Provinces of England*, recommending reform of local government and devolution of central government functions to provide a system of federal government with 12 provincial parliaments replacing the county system. The movement towards regionality was given further impetus by the work of G.D.H. Cole in *The Future of Local Government* (1921), in which he advocated dividing England into nine regions.

In 1969, in the context of a rising tide of Welsh and Scottish nationalism, a Royal Commission on the Constitution was established under the chairmanship of Lord Kilbrandon. This Commission published a majority report in 1973 proposing directly elected assemblies for Scotland and Wales and devolution of some powers to the English regions. The members were split on whether both legislative and executive powers should be devolved to the Scottish and Welsh Assemblies, but did agree that there should not be any devolution of legislative powers to England or to the English regions. The Commission was equally opposed to the introduction of a federal system.

**The New Planning Framework for the UK**

However, town and country planning in Britain is a very different activity today from that which emerged in the early years of the twentieth century, and it is no longer the preserve of local government. The 'Modernising Planning' initiative of the 1997 Labour Government has led to the creation of a new and more comprehensive planning framework for the UK, as illustrated in Table 3.2. It is a welcome move to enhance the statutory process in a European, national and strategic policy-making structure, while strengthening the system's role at the local level through strong and relevant development plans and a responsive and efficient planning control service.

**Table 3.2: The Planning Framework for the UK in 2000**

| SUPRA-NATIONAL | - EU<br>- ESDP<br>- Structural funds<br>- Regions |
|---|---|
| NATIONAL/UK | - PPGs<br>- RPG<br>- NSPF<br>- Regional policy |
| NATIONAL/REGIONAL | - Devolution for Wales, Scotland and Northern Ireland<br>- England and Wales share legislation<br>- Devolution for English regions? |
| REGIONAL | - English regions<br>- RDAs<br>- RPG<br>- London |
| LOCAL | - Local authorities: counties, districts and unitary authorities<br>- Concordats with DETR<br>- Bottom-up approach versus top-down<br>- Local governance and civic pride |

*Source:* Compiled by the author

Whether this more comprehensive UK planning system should be viewed as the product of a revolution or merely an evolution of a well-established system is open to debate. The policy statement 'Modernising Planning' (DETR, 1998) concluded that a modern planning system needs to build on the principles which were established over 50 years ago with the 1947 Act. Whilst people abroad refer to the 'British' planning system, and perhaps such a system existed with the 1947 Act, this is no longer the case. Since 1974 Scotland has worked within its own system of National Planning Policy Guidelines (NPPGs), and since 1996 Wales has worked within its own system of PPG. This has left England operating with its system of PPGs and RPGs. This fragmentation of the 'British' planning system might be expected to continue. Much will depend on the Government's future commitment to further devolve decision-making away from central government and towards the regions (Tewdwr-Jones, 1997; 1998).

This raises the question as to what are the alternative scenarios for the planning system in Great Britain/UK. Table 3.3 presents 10 scenarios for spatial planning in the UK. How far will the recent focus on 'regions and Europe' be taken? Does this necessarily diminish the role of national and local governments? It is still very early days in the lifetime of the new assemblies for Wales and Northern Ireland (on hold at the time of writing) and the Scottish Parliament. Much will depend on how far and fast they will exercise their new powers and functions, which are significant but limited.

The Royal Town Planning Institute initiated a project in 1998 to investigate the scope for a National Spatial Planning Framework (NSPF). This was reported, in its first stage, in Alden (1999) and other articles published in the same journal issue. One of the objectives of this initiative was to demonstrate which issues in planning require a national context for their resolution. As Table 3.4 illustrates, there is potentially a longer rather than shorter list of such issues.

Just as there are alternative scenarios for the future of the UK planning system, as illustrated in Table 3.3, so there are alternative institutional models for regional levels of governance and spatial planning. These are shown in Table 3.5. The UK exhibits examples of all the first five of the institutional models shown, but shows little capacity at present to embrace the sixth model of a federal system. Although Britain has had a very centralised planning system, recent changes have been significant in devolving decision-making to the regional level (Baker *et al.* 1999; Murdoch and Tewdwr-Jones, 1999).

**Table 3.3: Ten Scenaros for Spatial Planning in the UK**

| Scenario | Concept |
|---|---|
| 1 | Decentralise decision-making, especially at strategic level, particularly to the regions. |
| 2 | Decentralise decision-making and planning powers to the regions through more devolution. |
| 3 | Continue strong national planning policy guidance to establish and maintain national standards, and a framework for local government to promote consistency. |
| 4 | Discontinue national planning policy guidance as decision-making is decentralised to regions and localities, and the global framework is established by the European Spatial Development Perspective (ESDP). |
| 5 | Retain national planning policy guidance only for a small number of projects when decentralisation of decision-making is not possible. |
| 6 | Strengthen national planning policy guidance as a conduit to ESDP, and for regions and local governments to EU policies and strategies. |
| 7 | Retain all or some combination of the current spatial levels of planning, that is EU, national, regional and local. |
| 8 | Pursue United States of Europe with focus on Europe and regions and reduced role for national state and local authorities. |
| 9 | Continue and strengthen the local government-based plan-led system. |
| 10 | Encourage more bottom-up planning and less top-down planning through stronger local and community-based planning. |

*Source:* Adapted from Alden, 1999

**Table 3.4: Planning Issues Requiring a National Context and Perspective**

1. Providing a national strategic vision for the UK within a global economy: planning as an important element of this vision.
2. Joining up physical land-use planning with macro-economic policy, with planning system helping to deliver a competitive economy.
3. Giving an overarching view of 'national' plans being produced by Scotland, Wales, Northern Ireland and English regions.
4. Meeting the UK Government's international obligations.
5. Interfacing with EU spatial development policy and initiatives, and particularly access to Structural Funds.
6. Interfacing with EU spatial planning systems and policies, and particularly the ESDP.
7. Promoting and considering implications of global and trans-national/cross-border issues.
8. Promoting consistency in planning practice across the nation.
9. Establishing and promoting norms/standards of best planning practice across the nation.
10. Injecting spatial considerations into macro-economic and social policies of central government.
11. Providing a framework for local government and planning practitioners through national guidance, to promote certainty and support for planning decisions.
12. Promoting innovation and good practice in planning practice through advice notes.
13. Addressing strategic planning issues which contain important national dimensions, priorities and policy options.
14. Addressing major and infrastructure planning projects which can only be decided at a national level.

*Source:* Compiled by the author and adapted from Alden, 1999

**Table 3.5: Alternative Institutional Models for Regional Governance and Spatial Planning**

1.  Administrative decentralisation by central government, e.g. regional offices.

2.  Administrative devolution from central government, e.g. a Minister within each region, as in the past with Welsh Office and Scottish Office.

3.  *Ad hoc* regional institutions created by central government, e.g. Development Board for Rural Wales or Highlands and Islands Development Board in Scotland.

4.  Inter-local authority institutions, e.g. common action taken by local planning authorities in Scotland to produce Structure Plans.

5.  Regional or provincial government: a term used to embrace a wide range of institutions all characterised by regionally elected assemblies with either legislative or executive functions, or both.

6.  Federal model: transfer of sovereignty from national to regional parliaments for most activities, e.g. Germany, Belgium and Austria in EU, or USA, Australia, Canada, India, and so on.

*Source:* Compiled by the author and adapted from Alden, J. and Morgan, R., 1974

## Planning in Wales: Contrasts with Regional Planning Elsewhere in the UK and Ireland

The current system of regional planning in the UK is a considerable advance from hitherto. We now have assemblies for Wales and Northern Ireland, and a Parliament for Scotland. There are also Regional Development Agencies (RDAs) and RPGs for the English regions and a new planning authority for London with an elected mayor. This new system is producing some exciting examples of regional planning which focus upon more strategic thinking on the achievement of sustainable economic development and greater spatial equality in prosperity within regions. PPG 11 (DETR, 1999b) on Regional Planning has offered new guidance on best practice and strategic issues. Whilst neither Wales nor Scotland have produced real equivalents of RPG in England, the English regions, Northern Ireland, and Ireland all contain good examples of contemporary regional planning practice.

For example, the South West of England RPG provides a framework for the plans and strategic decisions of a range of public, private and voluntary sector agencies in relation to land use, transport, economic development and the environment over the period up to 2016 (South West Regional Planning Conference, 1999). The South West exhibits spatial differentials in economic prosperity similar to Wales, ranging from 70% of UK per capita Gross Domestic Product (GDP) for Objective 1 Cornwall to 114% for Gloucester (comparable to the Welsh Objective 1 area and Cardiff/South East Wales). The South West RPG contains a strategy for both its hot and cold spots. In contrast to Wales, the South West made smooth progress with its Objective 1 support, its Single Programming Document (SPD) and its spatial strategy. The lack of an adequate spatial planning perspective was one of the issues raised by the European Commission on the Welsh Objective 1 bid in 1999.

The RPG for the South East of England contains a core strategy for the region, a vision and key development principles (SERPLAN, 1998). It emphasises that the guidance must be taken into account by local planning authorities in preparing their development plans, and may be material to decisions on individual planning applications and appeals. The core strategy sets out the key components of the spatial structure of the region. It also assesses the region's role within the EU and UK.

In marked contrast to Wales, the East Midlands region has not had a strong regional identity. However, its Regional Assembly has produced regional perspectives and strategies on a number of issues, including an integrated regional strategy (East Midlands Regional Assembly, 2000). The work of its RDA and RPG matches that of any region in the UK and confirms a commitment to promoting the future of the East Midlands region. Other perspectives are being prepared on social inclusion, housing and rural development.

Ireland, with a population similar to Wales (i.e. 3.6 million and 2.9 million respectively), has stood out as a recent economic success story, based in part on its Structural Funds support. It has prepared a National Development Plan 2000-2006 which contains a regional development strategy (Department of Finance, 1999). This will guide public capital expenditure over the 2000-2006 period. The Plan consists of three Interregional Operational Programmes and two Regional Operational Programmes, one for each of the two new NUTS Level 2 regions created recently. In the Plan, the Government has confirmed its commitment to produce a National Spatial Policy for Ireland within two years.

Northern Ireland produced its draft regional strategic framework, 'Shaping our Future', in December 1998, and the Panel conducting the Public Examination published its report in February 2000. Again, the documents contained a context and guiding principles, and a spatial development strategy illustrated in the draft strategy document. The documents placed particular emphasis on economic growth and spatial development, and balanced growth. For many years Wales has been the poorest region in GB in terms of GDP per capita and Northern Ireland the poorest in the UK. However, Northern Ireland has now closed its prosperity gap in relation to the UK, unlike Wales. Moreover, Northern Ireland is poised to overtake Wales in the GDP per capita stakes, which will leave Wales the poorest 'region' in the UK. In 1966, Wales GDP per capita was 87.4% of the UK average and 83.1% in 1996, i.e. no improvement. In contrast, Northern Ireland's GDP per capita improved from 64.9% in 1966 to 81.2% by 1996, i.e. similar to Wales.

Wales now has a comprehensive planning framework, as illustrated in Table 3.6. It has had a special institutional framework for planning within the UK for many years, with the creation of the Welsh Development Agency (WDA) as long ago as 1975 (its RDA), and three National Park Authorities preparing development plans. In the absence of an all-Wales strategic planning policy, the WDA has published its strategic corporate plans.

**Table 3.6: The Welsh Planning System in 2000: A Comprehensive Planning Framework for Wales**

1. National Assembly for Wales (i.e. 'regional' government).

2. Since 1996 Wales prepared own PPG, i.e. Planning Policy Guidance (Wales) 1996, and 1999, revised.

3. PGW supplemented by Technical Advice Notes (TANs) on PPG-type issues.

4. Planning policy, which includes subordinate legislation.

5. A Planning Inspectorate which will continue to decide most appeals. NAW decide others.

6. Minerals Planning Guidance (MPG).

7. 22 unitary local planning authorities and 3 National Park authorities preparing development plans.

8. Welsh Development Agency (created 1975) adopts role of RDA in England; acquires powers of quangos including DBRW, CBDC, LAW. Published WDA Corporate Plan 2000–2003.

9. NAW published Objective 1 SPD to EU (1999) and EU response (1999), and NAW strategic plan 'Better Wales' (2000).

10. NAW created Land Use Planning Forum to prepare Planning Guidance Wales (PGW) 2000 (Revised). RTPI creates Welsh Planning Policy Panel.

11. NAW confirmed in May 2000 its intention to prepare a National Spatial Planning Framework for Wales, comparable to RPG in England. The NSPF to address spatial issues throughout Wales.

*Source:* Compiled by the author

The missing link in the planning system for Wales has been its spatial or regional perspective, i.e. a regional development strategy equivalent to RPG in the English regions. The National Assembly for Wales (NAW) has emphasised the importance of providing economic prosperity and sustainable development throughout Wales. Wales has 22 unitary planning authorities. Prior to the creation of the NAW, there was a consensus in Wales that the planning system lacked strategic and spatial guidance. The NAW is now to embark on filling this vacuum. The National Spatial Planning Framework will meet a number of urgent needs. It will provide a regional framework for the preparation of local authority development plans. It will incorporate a regional transport strategy and provide the longer-term planning framework for the strategy of the WDA. It is also important for land-use planning to be closely integrated with other strategies of the NAW, and particularly those of economic development and sustainable development. A Wales National Spatial Planning Framework will also enhance the European context facing Wales. It will improve access to Structural Funds, and help maintain consistency between the ESDP and a spatial perspective for Wales.

Much interest is currently focused in Wales on its Objective 1 programme of Structural Funds for 2000–2006. Its allocation of £1,173 million for West Wales and the South Wales Valleys is the largest sum allocated by the European Commission to the UK for Objective 1 areas (the others being Cornwall and the Isles of Scilly, Merseyside, and South Yorkshire). The Objective 1 status covers 15 of the 22 unitary authorities in Wales, illustrating the extent of low incomes and economic activity, covering 69% of the total population of Wales. Although the EU and Welsh matching funding of more than £2 billion appears a large sum, at £300 million a year it represents just 1.2% of Welsh annual GDP. The spending of these monies will need to be carried out in a strategic way and within a regional perspective, to achieve the objectives of the NAW, and in particular to close the Welsh prosperity gap with the other UK regions.

These selected examples of the RPG, or their equivalent, illustrate the current robust state of regional planning in the UK. However, where do we go from here? The English regions appear to be very productive in the new age of regional planning. Wales and Scotland, however, which have enjoyed a special institutional framework for planning since the establishment of their RDAs (WDA and Scottish Development Agency (SDA)) twenty-five years ago, plus a Secretary of State for each in the UK Cabinet, appear to have held back from embracing a more explicit form of spatial planning. The European dimension of planning is important for all the regions of the UK, whether they are core or periphery in economic prosperity terms. Current developments within the EU, and the recent experience of future applicant countries, suggest that regional planning and governance may become stronger rather than weaker in an enlarged EU.

## The Implications of the European Dimension of Planning for the UK

The European dimension of planning for the UK has increased in importance during recent years, and may be expected to continue to do so. Table 3.7 illustrates the 76 NUTS Level 1 regions of the EU. Agenda 2000 (Commission of the European Communities, 1997a) has examined the issues surrounding a stronger and wider EU. The Structural Funds programme of £200 billion for 2000–2006, together with the series of ESDPs culminating in the 1999 Potsdam final draft, represents an increasing focus on European spatial planning policy for the UK regions. The current 15 member countries of the EU have a combined population of 387 million, a size which requires a spatial development strategy. The regional dimension is an increasing focus for the EU because of the extent of regional economic disparities and the importance of addressing economic and social cohesion. It is also worth remembering that the 10 applicant countries of Central and Eastern Europe have a population of 104 million. During the next 10 years or so, the 25 member countries of the EU may have a combined population of nearly 500 million, outsizing by far the 270 million of the USA.

There are increasing examples of EU countries devolving power to their regions and localities, and becoming less of the classic unitary state model. Table 3.7 shows that with the UK devolving power to some of its regions, it is no longer the classic unitary state that it once was. Indeed, the three remaining classic unitary states are very small, i.e. Luxembourg, Greece and Ireland. The three federal states (Austria, Germany and Belgium) and two regionalised unitary states (Italy and Spain), i.e. categories D and E in Table 3.7, hold over 51% of the EU's population. If category C is added to this, i.e. unitary states devolving power to the regions, this figure grows to 75%. The classic unitary states now account for only 3.8% of the EU's population. The tendency, therefore, is for an increasing EU focus on the regional dimension.

Table 3.8 presents a typology of regional government in the 10 Central and Eastern European applicant countries as at 2000. Once centrally-planned economies, they have all produced new constitutions as a condition of eligibility for EU membership. Power is being devolved to regions and localities. These former centrally-planned economies have strong regional structures and much experience of spatial planning. Their combined population of 104 million, and their relatively low prosperity levels as illustrated by the GDP figures in Table 3.8, will produce increased momentum to engage the regional focus of both the Structural Funds and ESDP.

**Conclusions**

After nearly 100 years of statutory planning and interest in regionalism within the UK, the future outlook for the activity of regional planning appears robust. Pressures within both the UK and EU suggest that the regional dimension to governance and planning may be strengthened in the future. The membership of the 10 waiting applicant countries to the EU is likely to enhance the regional focus. There are, however, several alternative scenarios for the future direction of spatial planning in the UK, and a little uncertainty as to which will emerge as the most prominent. As illustrated in Table 3.2, the UK now has a new planning framework which looks very different to that of a decade ago. Whilst regional planning in the UK is an idea whose time may have come, the future pace of change is not at all easily predicted.

The pace of change is also difficult to predict in relation to EU enlargement, and particularly the membership of the Central and Eastern European applicant countries. The current 15 member countries will be looking anxiously at the future application and composition of EU regional policy and how it affects them. This comes at a time when decision-making on regional aid for both member countries and EU-wide is moving upwards towards Brussels rather than downwards to regions. The current interest in regionalism and spatial planning can be expected to gain even further momentum than it has to date.

## Figure 3.1: NUTS Level 1 Areas in the European Union

**Table 3.7: A Typology of Regional Government in Current 15 EU Countries**

| | |
|---|---|
| A. CLASSIC UNITARY STATES | LUXEMBOURG<br>GREECE<br>IRELAND<br>UK[1] |
| B. UNITARY STATES DEVOLVING POWER TO LOCAL AUHTORITIES | DENMARK<br>SWEDEN<br>FINLAND[2] |
| C. UNITARY STATES DEVOLVING POWER TO THE REGIONS | PORTUGAL<br>FRANCE<br>NETHERLANDS<br>UK<br>FINLAND |
| D. REGIONALISED UNITARY STATES | ITALY<br>SPAIN |
| E. FEDERAL STATES | AUSTRIA<br>GERMANY<br>BELGIUM |

[1] The UK was a classic unitary state until powers devolved to Wales, Scotland and Northern Ireland in 1999. It is now a devolving unitary state.
[2] Finland is in process of devolving powers to some of its regions.

*Sources:* Balchin, P., Sykora, L., and Bull, G., 1999
Commission of the European Communities, 1997b
*The Statesman's Yearbook 2000*

**Table 3.8: A Typology of Regional Government in 10 Central and Eastern Europe Applicant Countries of EU as at 2000**

|  |  | 2000 Population (million) | 1995[1] GDP/head as % EU average |
|---|---|---|---|
| A. CLASSIC UNITARY STATES | New constitutions for all 10 applicant countries since 1990: now transitional between typology A and B/C |  |  |
| B. UNITARY STATES DEVOLVING POWER TO LOCAL AUTHORITIES | Hungary<br>Poland<br>Romania<br>Latvia<br>Slovenia<br>Estonia<br>Lithuania | 9.81<br>38.73<br>22.50<br>2.40<br>1.99<br>1.42<br>3.69 | 37<br>31<br>23<br>18<br>59<br>23<br>24 |
| C. UNITARY STATES DEVOLVING POWER TO THE REGIONS | Poland<br>Slovak Republic<br>Bulgaria<br>Czech Republic | As above<br>5.37<br>8.31<br>10.19 | As above<br>41<br>24<br>55 |
| D. REGIONALISED UNITARY STATES |  |  |  |
| E. FEDERAL STATES |  |  |  |

[1] GDP at purchasing power standards: if GDP Market Price Latvia = 8%

*Sources:* Commission of the European Communities, 1997b
*The Statesman's Yearbook 2000*

# References

Alden, J.D. (1999), 'Scenarios for the Future of the British Planning System: The Need for a National Spatial Planning Framework', *Town Planning Review*, vol. 70(30), pp. 385–407.

Alden, J.D. and Boland, P. (eds.) (1996), *Regional Development Strategies: A European Perspective*, Jessica Kingsley, London.

Alden, J.D. and Morgan, R. H. (1974), *Regional Planning: A Comprehensive View*, Leonard Hill Books, Pitman Press.

Baker, M., Deas, I. and Wong, C. (1999), 'Obscure ritual or administrative luxury? Integrating strategic planning and regional development', *Environment and Planning B*, vol. 26, pp. 763–82.

Balchin, P., Sykora, L. and Bull, G. (1999), *Regional Policy and Planning in Europe*, Routledge, London and New York.

Cole, G. D. H. (1921), *The Future of Local Government*, Cassell, London.

Commission of the European Communities (1997a), Agenda 2000: For a Stronger and Wider Union, *Bulletin of the EU*, Supplement 5/97, Luxembourg.

Commission of the European Communities (1997b), *The EU Compendium of Spatial Planning Systems and Policies*, CEC, Brussels.

Commission of the European Communities (1998), Reform of the Structural Funds, *Newsletter No. 50*, March 1998.

Commission of the European Communities (1999), *European Spatial Development Perspective Towards a Balanced and Sustainable Development of the Territory of the EU*, Office for Official Publications of the European Communities, Luxembourg.

Department of the Environment, Transport and Regions (1998), *Modernising Planning: A Policy Statement by the Minister*, January.

Department of the Environment, Transport and Regions (1999a), *Modernising Planning: A Progress Report*, DETR, London.

Department of the Environment, Transport and Regions (1999b), *Planning Policy Guidance Note 11. Regional Planning:* Public consultation draft.

Department of the Environment for Northern Ireland, (1998) *Shaping our Future. Draft Regional Strategic Framework for Northern Ireland*, DOE (NI), Belfast.

Department of Finance (1999), *National Development Plan 2000-2006*, Department of Finance, Dublin.

East Midlands Regional Assembly (2000), *The English East Midlands Integrated Regional Strategy*, East Midlands Regional Assembly.

Elliott, A. and Forbes, J. (2000), *Report of the Panel of the Public Examination on Draft Regional Strategic Framework for Northern Ireland*, DoE (NI), Belfast.

Fawcett, C. (1919), *Provinces of England*, Williams and Norgate, London.

Geddes, P. (1915), *Cities in Evolution*, Williams and Norgate, London.

Murdoch, J. and Tewdwr-Jones, M. (1999), 'Planning and the English Regions: Conflict and Convergence Amongst the Institutions of Regional Governance', *Environment and Planning C, Government and Policy*, vol. 17 (6), pp. 715-29.

Royal Commission on the Constitution 1969-73, vol. 1 Report, Kilbrandon Committee, Cmnd. 5460, 1973.

Samuel, W. (1999), 'Virtues and Vices of Regional Plans', *Planning*, 25 June 1999, p. 15.

Sanders, W. (1905), *Municipalization by Provinces*, Fabian Tract 125, Fabian Society, London.

SERPLAN (London and South East Regional Planning Conference) (1998), *A Sustainable Development Strategy for the South East*, SERPLAN, London.

South West Regional Planning Conference (1999), *Draft Regional Planning Guidance for the South West*, Government Office for the South West, Bristol.

*The Statesman's Year Book 2000*, Macmillan, Basingstoke.

Tewdwr-Jones, M. (1997), 'Plans, Policies and Inter-Government Relations: Assessing the Role of National Planning Guidance in England and Wales', *Urban Studies*, vol. 34(1), pp. 142-62.

Tewdwr-Jones, M. (1998), 'Planning Modernised?', *Journal of Planning and Environmental Law*, pp. 519-21.

Welsh Office (1996), *Planning Guidance (Wales): Planning Policy*, May 1996, Welsh Office, Cardiff.

Welsh Office (1999), *Planning Guidance (Wales): Planning Policy: First Revision*, April 1999, Welsh Office, Cardiff.

# 4 The Regional Agenda, Planning and Development in Scotland

GREG LLOYD

## Introduction

In the United Kingdom (UK), the prevailing political programme of the Labour administration centres on the widening and deepening of democratic processes as the principal key to renewing contemporary civil society (Giddens, 1998). The modernisation of the institutions and processes associated with social democracy has involved the devolution and decentralisation of power in order to re-enfranchise individuals, communities and localities and encourage active political participation. Devolution has involved the transfer of power from one elected body to another with responsibilities defined appropriately for scale, jurisdiction and geography. In effect, it involves the establishment of quasi-autonomous government in regions that have what may be considered to be a national cultural identity. However, this does not necessarily involve the transferring of full legal sovereignty (McNaughton, 1998). In practice, devolution has involved the transfer of specific legislative, executive and financial powers from Westminster to elected bodies in Scotland, Wales and Northern Ireland. The elected bodies remain subordinate to a legal framework defined by the centre (Bogdanor, 1999). This represents a process of constitutional reform which preserves the supremacy of Parliament, whilst giving identity to the former regions of a centralised entity and avoiding thereby a federalist administrative arrangement.

In practice, devolution is taking place in a differential manner throughout England, Wales, Scotland and Northern Ireland. This reflects a number of inherited and emerging political, electoral, cultural circumstances (Bogdanor, 1999). This is evident with respect to a range of issues including, for example, relative economic performance, industrial and corporate restructuring, social exclusion, urban-rural relationships and outcomes associated with inward investment. In part, the differential nature of devolution reflects established differences in institutional infrastructure and institutional capacity between the constituent parts of the UK. The uneven geographical coverage and institutional capacity of individual regional development agencies illustrates this point. It has been argued that devolution and the uneven institutional circumstances in which it is taking place are likely to result in varying lower-order or sub-national arrangements for strategic spatial planning, land use planning and institutional innovation (Lloyd and Tewdwr-Jones, 1997).

The processes are now well underway. Different forms of spatial planning have emerged in the context of the different constitutional arrangements. In Wales and Northern Ireland, for example, a relatively assertive approach is beginning to establish itself, which is based on the preparation of a national economic development strategy. This strategy sets out a framework for economic management, containing a spatial dimension and institutional innovation at its core. In Northern Ireland, a draft regional strategic framework asserts that it 'is not a fixed blueprint or master plan. Rather it is a framework, prepared in close consultation with the community, which defines a vision for the region and frames an agenda which will lead to its achievement' (Northern Ireland Office, 1999, p. 2). In both Wales and Northern Ireland, a strategic and spatial dimension is linked relatively explicitly to institutional innovation and the preparation of strategic planning guidance. In England, there is an attempt at institutional capacity-building through the introduction of the Regional Development Agencies (RDAs) which operate within a contested and congested institutional landscape, but which are an attempt to enhance intellectual and social capital in the individual regions (Roberts and Lloyd, 2000).

In Scotland, the Parliament and Executive is in the process of devising its own approach to forms of strategic planning, institutional innovation and economic development. This paper considers the evolving nature of the Scottish arrangements in order to provide a framework for accommodating interests associated with development at national, regional and local levels of decision making in the new devolved geography of power.

## Strategic Spatial Planning Innovation and Constitutional Reform

A case for strategic planning rests on the arguments for enhanced institutional capacity for governance, planning and public administration which can set a basis for consensus and a framework for consistent policy implementation (Healey, 1997). In this context (Healey, 1997, p. 21), spatial planning may be described as 'a social practice through which those concerned with the qualities of places and the spatial organisation of urban regions collaborate to produce strategies, policies and plans to help guide specific decisions in order to regulate and invest in development activity'. It is significant that a principal focus for a spatial planning process is that of attaining balance in terms of social, economic and environmental development and change within a national or regional economy (Alden, 1996). This rejects the neo-liberal perspective that uneven economic performance (and allied social and environmental effects) is intrinsic to the necessary functioning of capital and labour markets.

Spatial planning seeks to provide a management framework that can facilitate and regulate economic activity in the wider social and community interest. This involves the management of the benefits and costs of economic change and restructuring. In this respect, the potential of enhanced institutional capacity is important in establishing the structural capacity for, and the cultural means of, securing effective and accountable decision-making at a time of change. The capacity of the new constitutional arrangements is also significant in terms of the need for appropriate economic, social and environmental strategies. Specifically, an enhancement of institutional capacity in relation to spatial planning may allow more effective engagement with the geopolitical division and variations within national economies.

The creation of enhanced institutional capacity for strategic spatial planning can help to realise the full potential of the participative and integrative processes involved in land use planning. Indeed, as Healey (1997, p. 18) has indicated, 'strategic spatial planning can provide a transparent, fair and legitimate way of recognising and responding to the multiplicity of stakeholders, interests and value conflicts that arise in urban regions'. This can constitute a process of institutional innovation which, as Motte (1997) has indicated, can occur in the context of spatial planning where new spatial planning systems are applied, where such systems are applied in a different way from in the past, or where new mechanisms, processes or organisations are created to allow planning systems to work more efficiently. This would embrace the differential approaches in Scotland, England, Wales and Northern Ireland as sub-national responses to localised circumstances and performance.

Spatial planning involves a number of inter-related concepts (Healey, 1997). These include the structure-agency nexus in terms of the relationship between contextual conditions and the construction of behaviour and response to those changing circumstances; social interaction in generating the conditions for the accumulation of knowledge and value in society; the social and political context in which key players operate; processes of social and political mobilisation associated with spatial planning; policy discourse in determining common agendas for action or negotiated settlements; and the core elements of institutional capital and institutional capacity (Healey, 1997). In terms of its functions, spatial planning creates benefits for the management of economic activity, social and community regeneration and environmental sustainability. It establishes a framework in which common problems can be identified, available resources marshalled, collective priorities determined and policies devised. In addition, spatial planning attempts to address the dynamics of urban and regional change by mobilising political and empowering forces; it then attempts to secure common agendas for action which reconcile the needs, expectations and demands of communities with available resources, space and place. In this way, spatial planning processes can capture the social and political dynamics of communities (empowerment), and can provide a framework for investment and development (governance).

Central to the notion of spatial planning is that of institutional innovation, in terms of institutional capital and capacity (Amin and Thrift, 1995). Just as economic activity and social opportunity are not even over time and space, so too institutional provision for the management of change is uneven. Reference can be made in this context to the differential provision of regional development agencies across the UK. Institutional innovation is important, therefore, in providing the mechanisms appropriate to facilitating a spatial planning impetus. Institutional innovation comprises two elements: firstly, notions of institutional capital and secondly, institutional capacity (Healey, 1997). Institutional capital can include the components of intellectual, social and political capital. Together, these comprise the institutional capacity of a locality to understand the nature of regional and local dynamics and to be able to respond to change. The concept of institutional capacity 'refers not merely to the formal organisational structures, procedures and policy measures that actors can draw upon in developing strategic spatial plan making activity. It also includes the collective store of relationships and alliances and the institutional loci of interactions' (Healey, 1997, p. 23). It can be viewed as the way in which an institutional infrastructure can manage change in the wider social or community interest. The specific questions for this paper are as follows: to what extent is institutional capacity building taking place within the contextual framework of devolution? and what form does any institutional capacity building take in terms of enhancing or innovating with respect to intellectual, social and political capital?

This concept sits comfortably alongside processes of devolution and decentralisation of arrangements for economic development and innovation (Cooke and Morgan, 1998). It is very much a progression beyond the traditional forms of state intervention (social democratic) and neo-liberal approaches. It also gels with the political ideas associated with the Third Way agenda. Traditional institutional development was focused on service delivery, such as training and technical assistance. This approach represented a combination of the traditional top-heavy state-led intervention and latterly the neo-liberal agendas of service delivery at minimal cost. Now, however, there is greater emphasis on process, the arrangements for governance, empowerment and capacity development. The contemporary emphasis on institutional innovation attempts to correct the balance away from an exclusively supply-side domain. It may be seen as a synthesis of the managerial, political and economic elements of institutional development. It sits comfortably alongside the new approaches to private and public innovation in facilitating regional development, and it stresses tangible and intangible processes of negotiation, dialogue and consensus building. In short, it reflects and captures many of the ideas associated with the spatial planning repertoire.

**Regionalism in Scotland**

In Scotland, devolution has involved the creation of a Scottish Parliament and Scottish Executive, which have assumed responsibilities for a range of administrative, political and financial functions. There has been the contingent delegation of other responsibilities, including local government, housing, social exclusion, urban regeneration, land use planning, environmental protection and management, which have then passed to the new institutions (Himsworth and Munro, 1998). However, Westminster has retained legislative supremacy over reserved functions, including economic policy, foreign affairs and defence. However, Hassan (1999, p. 16) argues that 'regulatory policies offer a potential model for the Scottish Parliament. At one level, they do not involve Government directly running services or setting up new bureaucracies and expenditures, but involve them in setting agreed rules that are then implemented by other agencies. Active government can then be linked with the politics of decentralisation. Regulatory policies are thus an appropriate vision of government in an age where on one level the state is under attack and critique, yet more and more is expected from it by the general public'. For the purposes of this paper, a number of innovations may be cited to illustrate the changing nature of institutional capacity and innovation in Scotland.

Firstly, an overarching theme in the Parliament's relatively short tenure has been that of an emphasis on the issues associated with social inclusion and social justice. This reflected the prevailing social and community circumstances together with the inherited policy regime put in place to address the geography of exclusion and disadvantage in Scotland. The Scottish Parliament re-defined this policy agenda and sought to consolidate and extend it with the principal aim of achieving greater integrated local action to achieve the measurable outcomes associated with social justice. Specifically, it is stated that 'integrated local actions, often involving expenditures by different government agencies and a range of government departments, are now essential aspects of partnerships to promote area regeneration and individual well-being' (Scottish Executive, 2000a, p. 19). This reflects the importance of partnership working in Scotland which itself reflects an established corporatism in public affairs. It has been argued, for example, that there is evidence 'to support the view that, although there may have been a rejection of corporatist or consensual politics south of the border, in many respects they have survived in some areas of policy in Scotland' (Brown, McCrone and Paterson, 1996, p. 106).

The contemporary policy measure for urban regeneration, for example, reflects the view that 'Scottish circumstances differ from England in that those suffering exclusion in Scotland are disproportionately concentrated in specific communities, and there has been more experience of effective urban regeneration policies originally pioneered in Scotland and maintained in later years by local government and others' (Dewar, 1998, p. 1). Urban regeneration policy is predicated on the most needy members of society; the co-ordination and filling in of gaps between existing programmes in order to promote inclusion; and, an attempt to prevent people becoming socially excluded. Partnership working is intended to demonstrate pro-active working, synergy between different organisations, a widening of the potential for leverage in funding, co-operation and co-ordination of activity, strategic and long-term thinking and the implementation of flexible, innovative and responsive policies to local need (Chapman, 1998). Social justice operates principally at the local level – either in geographically defined areas or in thematically defined social groups. The emphasis on competitive bidding for available resources (Turok and Hopkins, 1998) constrains the emergence of a strategic approach to policy and planning. Social justice has, however, influenced the thinking of the Parliament and, as a consequence, there is an acknowledgement of the need for managed intervention. This is reflected in the more recent economic initiatives.

Secondly, a stronger economic (if not explicitly economically deterministic) dimension has emerged through the deliberations of the Parliament's Enterprise and Lifelong Learning Committee concerning the provision of local economic development services agendas. The Committee examined the overall arrangements for the delivery of local economic development, post-school vocational education and training together with business support services in Scotland. It concluded that there is considerable congestion in these arrangements, with evidence of confusion, overlap, duplication and inter-organisational competition for available resources (Enterprise and Lifelong Learning Committee, 2000). The Committee confirmed, however, the extent to which positive partnership working is being achieved in individual localities with positive results, but advocated that the roles of the key players have to be more precisely defined to secure a more productive deployment for available resources. Indeed, the Committee suggested that enhanced partnership working at the local level would not achieve the broader strategic objectives of national and regional economic development. In short, Scotland's institutional capacity requires revision.

The Enterprise and Lifelong Learning Committee (2000) advocated an economic development strategy for Scotland based on growth through competitiveness and sustainability, but taking into account regional development and social integration. To facilitate this national perspective, it advocated what may be considered an implicit spatial or sub-national (regional) dimension. The Committee argued for the establishment of local economic forums for each Local Enterprise Company (LEC) area of the Scottish Enterprise and Highlands and Islands Enterprise networks. Each local economic forum would be based on local major interests, including the local authorities, LEC, Chambers of Commerce and institutions of local, higher and further education. Each forum would create a local economic strategy for its area including the unambiguous delineation of the responsibilities of each organisation involved in local economic development (Enterprise and Lifelong Learning Committee, 2000).

This line of reasoning has been followed by the subsequent publication of the economic development framework for Scotland (Scottish Executive, 2000b). This set out a vision for the Scottish economy, which is based on attaining economic growth through international competitiveness, regional development across Scotland, social integration and the principles of sustainability. Echoing the approach of the Enterprise and Lifelong Learning Committee, the economic development framework seeks to most precisely define the roles and priorities of the bodies involved to enable more effective working. The economic development strategy is predicated on the need for partnership, comprehensiveness and the adoption of a longer-term perspective. Attention is drawn, for example, to the need to 'consider the basic organisation of economic activity and the degree to which it promotes entrepreneurial dynamism. The environment must facilitate a range of activities including the establishment of new enterprises, the establishment of productive bases in Scotland by overseas enterprises, the expansion of small enterprises, collaboration and joint ventures between productive enterprises and centres of knowledge and research, and the development of the formal and informal networks that help to lower the costs of economic transactions' (Scottish Executive, 2000b, 4.1). This suggests particular roles and priorities within the Scottish economic environment.

Finally, there has been a growing interest in the processes associated with community planning in Scotland. This involves local authorities working together with their principal public sector partners to plan for and deliver services that meet the needs of their local constituent communities. More formally, it has been described as a means by which local authorities can put forward an informed view of the challenges and opportunities facing their communities (Community Planning Working Group, 1998). This process is intended to be inclusive of the key players and interest groups involved in those localities so as to facilitate the overall well-being of the community at large (Sinclair, 1997). It was recently announced, for example, that the Scottish Executive would introduce a statutory power of community initiative and of community planning in a forthcoming local government bill. It was asserted that the new power of community initiative and community planning 'will help make a reality of joint working with other bodies; cross cutting initiatives and it will provide strong foundations for community planning. In essence, it will set a framework within which Councils could truly embrace a community leadership role, and so help deliver renewal of communities across Scotland' (Alexander, 2000). Community planning illustrates the nature of institutional innovation in Scotland to date and is an example of an attempt to enhance existing institutional capacity to address prevailing sub-national planning issues.

**Community Planning in Scotland**

In July 1997, the Secretary of State for Scotland and the Convention of Scottish Local Authorities (COSLA) established a joint working group – the Community Planning Working Group – that was comprised of officials from the two organisations. Its responsibility was to examine the ways in which Scottish local authorities engage in partnerships with other bodies to provide for and promote the economic, social and environmental well-being of the communities they serve. In this sense, the Community Planning Working Group is performing as an institutional sponsor in seeking the legitimacy and continuity in the social construction of community planning and leadership by local authorities (Hannigan, 1995). The role of the Community Planning Working Group in this context is therefore important in discharging this role and laying the foundations for community planning in Scotland in popularising and advocating its adoption to address the perceived issues associated with the emerging agenda for local governance.

The Community Planning Working Group (1998) highlighted the leadership role of local government in asserting local agendas for the delivery of services and in providing a managed framework for effective inter-organisational relations. Indeed, community planning is presented as a process through which a Council comes together with other organisations to plan, provide for or promote the well-being of communities they serve. This can potentially operate across an entire Council area and at the level of local communities within a given Council area.

In particular, the Community Planning Working Group (1998) described the aims of community planning as:

- to improve the service provided by Councils and their public sector partners to the public through closer, more co-ordinated working;
- to provide a process through which Councils and their public sector partners, in consultation with the voluntary and private sector, and the community, can agree both a strategic vision for the area and the action which each of the partners will take in pursuit of that vision; and,
- to help Councils and their public sector partners collectively to identify the needs and views of individuals and communities and to assess how they can best be delivered and addressed.

A central purpose of the community planning concept is to address the potential and existing gaps in the existing array of statutory and non-statutory planning institutions and processes. The extent to which there is fragmentation in the institutional landscape in Scotland is shown in the results of a recent survey of the direct economic development activity undertaken by local government (Ekos, 2000). In this context, direct economic development includes business competitiveness, physical business infrastructure, training and human resource development, tourism and economic inclusion, community economic development. There is considerable diversity in the approaches adopted by individual Councils. It shows the extent to which Councils work with a wide range of public, private and voluntary organisations to deliver economic development services. In particular, joint working between Councils and other organisations has increased significantly since 1996. The reasons are perhaps to be expected and include the emergence of shared economic development strategies, participation in European Structural Fund programmes and emerging social inclusion agendas (Ekos, 2000).

The Community Planning Working Group (1998) argued that there is a need to ensure greater co-ordination between existing institutions, the associated processes and the communities for which they are intended. It concluded 'that there is at present a lack of structured overview across the various agencies about how they collectively could best promote the well-being of their communities. The kaleidoscope of local and subject-specific plans and partnerships was not related to a consistent attempt to develop a shared strategic vision for an area and a statement of common purpose in pursuit of that vision. Neither was it always clear who had the lead in developing this shared strategic vision' (The Community Planning Working Group, 1998, p. 3).

At a Council area level, the purpose of the community planning process would be to present an informed view of the challenges and opportunities facing the geographical communities and the different communities of interest. A community plan is envisaged as holding for between 5–10 years and is subject to annual review with clear statements of progress to the commonly agreed agenda for action. It would involve full consultation with individuals, communities and the private sector although it is clearly driven from the public sector community. The community planning process is also envisaged at working at more local levels of interest. It is clear that the concept of community planning involves a structure (the Plan) and a process of negotiation whereby the different interests and policy positions of all the bodies concerned with community are drawn together into a common agenda.

The Community Planning Working Group recommended that Councils and their partners should aim to produce their first community plans by the end of September 1999. Five local authorities were invited by the Scottish Office to act as pilots for the initiative and to draw up draft community plans by the end of 1998. The five Pathfinder local authorities were Highland Council, City of Edinburgh Council, Perth and Kinross Council, South Lanarkshire Council and Stirling Council. On the basis of this initiative there was general support for the concept of community planning from those organisations involved in the Pathfinder partnerships and the researchers concluded that the value of the process is beyond doubt and that there was a high level of agreement about the fundamental value of community planning within local councils and their partner organisations (Rogers *et al.*, 1999).

**Towards a Renaissance in Scottish Regionalism?**

In a devolved Scotland, the region becomes the nation and then sub-national or regional economic issues assume greater importance on political agendas. It is now acknowledged that Scotland comprises a series of inter-locking regional economies (Peat and Boyle, 1999). These are characterised by differential regional economic performances, change and the management of change. The indicators reflect many of the traditional characteristics of uneven economic performance, including differential unemployment, employment, Gross Domestic Product per capita indicators, structural imbalances and uneven patterns of inward investment and innovation. The regional imbalances represent a pressing issue for the Scottish Parliament and Scottish Executive as was acknowledged by the deliberations of the Parliament's Enterprise and Lifelong Learning Committee (2000). In advocating a national economic development strategy for Scotland, for example, the Committee explicitly noted the need for a regional development agenda. This was echoed in the subsequent economic development framework which would provide an agenda for strategic action as proposed by the Scottish Executive.

Traditionally, the Scottish Office (a role now carried forward by the Scottish Executive) has played a key role in the administration and governance of Scotland over time. It has established its own very distinctive identity and a distinguishing feature of the institution is the extent to which its activities are co-ordinated rather than being centrally determined (Parry, 1992). The Scottish Office has played a direct role in attempts to restructure the Scottish economy through public sector policy, funding and institutional activity in the immediate post-war period (McCrone, 1992; Danson, 1997). It has done this through the political legitimisation of corporatism articulated through exhortations of the national interest (McCrone, 1992). Until recently, strategic regional planning in Scotland has been enhanced by specific institutional arrangements. Regional development agencies, for example, represent a distinctive feature of Scottish public administration and which may be identified as having played an important role in the process of planning and economic development (Danson, Lloyd and Newlands, 1993). In local government, regional authorities played an important role. Strathclyde Region, for example, 'has been a unique experiment in regional governance; the only metropolitan regional council ever introduced in the UK and with an exceptional range of services by which to support its strategies' (Wannop, 1995, p. 113). In terms of metropolitan planning, Edinburgh, for example, has faced and continues to face development pressures associated with its role as capital city and financial centre of Scotland tempered by its historical advocacy of preservation of its architectural heritage and built environment (Hague and Thomas, 1997). In West Central Scotland, regional planning has reflected an entrenched industrial past and an underlying theme of collective public action to address the severe problems it faced with respect to housing and economic renewal (Wannop, 1986). More recently, these circumstances have changed.

Despite this rich history, tradition and experience, there is a strategic planning vacuum in Scotland (Lloyd, 1999). This is in spite of the fact that at the national level, the Scottish Office provided an overview of the management of the Scottish economy and its physical infrastructure with a strong regional emphasis involving the integration of economic and land use matters (Paterson, 1994). Today, the lack of an explicit economic planning agenda and the emphasis on specific land use issues has served to deflect attention away from strategic thinking. The political antipathy to regional planning contributed in no small measure to this. The experience of regional planning in Britain has always been a relatively chequered one (Glasson, 1992). Political preferences for the efficiency and effectiveness of intervention and markets has undermined the development of regional planning, which as a result, has been described as representing 'an enduring but inconstant feature of public affairs' in post-war Britain (Wannop and Cherry, 1994, p. 52). Local government re-organisation in Scotland has had an adverse effect on the regional arrangements by abolishing the regional councils and introducing single tier, all purpose unitary authorities (Midwinter, 1995; Hayton, 1994). The growth of quasi-government has also moved away from the idea of strategic thinking. This reflects 'a growth in central state powers and an attack on other sources of power held by local government and key policy areas in Scotland, together with an increased role given to the Scottish Office in improving central government policies in Scottish Society' (Jordan, 1994, p. 93). Quasi-government involves the transfer of power from the directly elected and relatively transparent system of central and local government to the closed world of quangos (Skelcher, 1998). There are a number of motives for deploying quangos in the structures of government and the processes of governance (Parry, 1999). However, there are issues of accountability and institutional fragmentation involved which undermine the notion of strategic thinking.

Spatial planning is described as 'a social practice through which those concerned with the qualities of places and the spatial organisation of urban regions collaborate to produce strategies, policies and plans to help guide specific decisions in order to regulate and invest in development activity' (Healey, 1997, p. 21). Deploying different language, the Town and Country Planning Association (1993) has suggested the case for regional planning based on an organisational structure which emphasises the interdependence of sectoral actions; allows for a comprehensive view of change and the management of change to be provided; provides a strategic vision or visions for a region; reflects both top-down and bottom-up priorities; be robust enough to cope with changing circumstances; reflects the views of the full range of regional stakeholders and allows for adjustments to be made to policy and its implementation. These features have not existed in Scotland in recent times. Moreover, the emerging economic development framework would not appear to satisfy these prerequisites fully. The community planning initiative does go some way towards them.

Community planning, although still in its infancy as an established aspect of public administration and partnership building, does attempt to create a framework in which common problems can be identified, available resources marshalled, collective priorities determined and policies devised. Community planning is the nearest process to strategic integration and rationalisation of activity in Scotland at the present time. There is no middle strategic regional ground or conduit to facilitate a sub-national agenda. Thus, neither the nationally based economic development framework nor community planning can address the dynamics of urban and regional change. There is no means as yet of mobilising strategic political and empowering forces or securing common agendas for action which are based on strategic jurisdictions. Another problem is that community planning as it has thus far evolved is aspatial. It is effectively divorced from the statutory land use planning system; as such it cannot provide tangible guidance for public and private sector decision makers. This would suggest that some attention must be paid to the integrated relationship between community planning and the community plans and structure plans, which can provide a land use, spatial and relatively strategic context. This is an area that requires careful attention in the near future. Indeed, it begs a deeper issue of where structure plans fit into the proposed economic development framework as a whole.

Significantly, the Enterprise and Lifelong Learning Committee (2000) has asserted that the local economic strategy which would be the outcome of each local economic forum should then become the economic element of the appropriate community plan. This is a bridging link, but given that there will be in excess of 20 fora and that several of these include a number of community plans, this represents a major challenge for the process. Neither the economic development framework, the proposed local economic fora nor community planning represent a sufficient degree of institutional innovation appropriate to facilitating a spatial planning impetus. There is little in terms of creating institutional capital, and community planning is a relatively limited attempt to enhance institutional capacity. At the local level, however, community planning does involve the characteristics of associational economics in terms of involving negotiation, dialogue and consensus building. It clearly has a role to play in the emerging arrangements for economic development.

## Conclusion

In practice, community planning is the most advanced element of the modernisation agenda in Scotland. The economic development framework is now seeking to establish new arrangements and new roles for economic and spatial management. The McIntosh Commission (1998), for example, advocated that local authorities be granted a power of general competence within which community planning would play a central role. This would certainly allow action to be taken by local authorities without the need for statutory powers and give councils more freedom to work with the other partners involved. This would be consistent with emerging thinking. There is a political intention to create such a competence through the community initiative which includes the processes of community planning (Alexander, 2000). However, many more questions remain to be answered before community planning plays its full role in regional planning frameworks in Scotland.

## References

Alden, J. (1996), 'Regional development strategies in the EU: Europe 2000+', in J. Alden and P. Boland (eds.), *Regional Development Strategies: A European Perspective*, Jessica Kingsley, London, pp. 1–13.

Alexander, W. (2000), *Local Government Modernisation*, Ministerial Statement, Edinburgh, June 8.

Amin, A. and Thrift, N. (1995), 'Globalisation, institutional thickness and the local economy', in P. Healey, S. Cameron, S. Davoudi, S. Graham and A. Madanipour (eds.), *Managing Cities: the New Urban Context*, John Wiley, London, pp. 91–108.

Bogdanor, V. (1999), *Devolution in the United Kingdom*, Oxford University Press, Oxford.

Brown, A., McCrone, D. and Paterson, L. (1996), *Politics and Society in Scotland*, Macmillan, London.

Chapman, M. (1998), *Effective Partnership Working*, Good Practice Note 1, Scottish Office, Edinburgh.

Community Planning Working Group (1998), *Report of the Community Planning Group*, Scottish Office, Edinburgh, July.

Cooke, P. and Morgan, K. (1998), *The Associational Economy, Firms, Regions & Innovation*, Oxford University Press, Oxford.

Danson, M.W. (1997), 'Scotland and Wales in Europe', in R. MacDonald and H. Thomas (eds.) *Nationality and Planning in Scotland and Wales*, University of Wales Press, Cardiff, pp. 14–31.

Danson, M., Lloyd, M.G. and Newlands, D. (1993), 'The role of development agencies in regional economic development', in M. Hart and R. Harrison (eds.), *Spatial Policy in a Divided Nation*, Jessica Kingsley, London, pp. 162–75.

Dewar, D. (1998), *The Scottish Social Inclusion Strategy*, Castlemilk.

Ekos (2000), *Survey of Economic Development Activity of Scottish Councils*, COSLA, Edinburgh.

Enterprise and Lifelong Learning Committee (2000), *Inquiry into the Delivery of Local Economic Development Services in Scotland*, First Report 2000, Edinburgh, May.

Giddens, A. (1998), *The Third Way. The Renewal of Social Democracy*, The Polity Press, Cambridge.

Glasson, J. (1992), 'The fall and rise of regional planning in the economically advanced nations', *Urban Studies*, vol. 29, pp. 505–31.

Hague, C. and Thomas, H. (1997), 'Planning capital cities: Edinburgh and Cardiff compared', in R. MacDonald and H. Thomas (eds), *Nationality and Planning in Scotland and Wales*, University of Wales Press, Cardiff, pp. 133–58.

Hannigan, J.A. (1995), *Environmental Sociology. A Social Constructionist Perspective*, Routledge, London.

Hassan, G. (1999), 'The new Scottish politics', in Hassan, G. (ed.), *A Guide to the Scottish Parliament*, The Stationery Office, Edinburgh, pp. 9–19.

Hayton, K. (1994), 'Planning and Scottish local government reform', *Planning Practice and Research*, vol. 9 (1), pp. 55–62.

Healey, P. (1997), *Collaborative Planning. Shaping Places in Fragmented Societies*, Macmillan, London.

Himsworth, C.M.G. and Munro, C.R. (1998), *Devolution and the Scotland Bill*, W. Green/Sweet and Maxwell, Edinburgh.

Jordan, G. (1994), *The British Administrative System. Principles versus Practices*, Routledge, London.

Lloyd, M.G. (1997), 'Structure and culture: regional planning and institutional innovation in Scotland', in R. MacDonald and H. Thomas (eds.), *Nationality and Planning in Scotland and Wales*, University of Wales Press, Cardiff, pp. 113–32.

Lloyd, M.G. (1999), 'The Scottish Parliament and the planning system: addressing the strategic deficit through spatial planning', in J. McCarthy and D. Newlands (eds.), *Governing Scotland: Problems and Prospects*, Ashgate, Aldershot, pp. 121–34.

Lloyd, M.G. and Tewdwr-Jones, M. (1997), 'Unfinished business? Planning for devolution and constitutional reform', *Town and Country Planning*, vol. 66 (11), pp. 302–04.

McCrone, D. (1992), *Understanding Scotland. The Sociology of a Stateless Nation*, Routledge, London.

McIntosh Commission, (1998) *Local Government and the Scottish Parliament*, Edinburgh.

McNaughton, N. (1998), *Local and Regional Government in Britain*, Hodder and Stoughton, London.

Midwinter, A. (1995), *Local Government in Scotland. Reform or Decline?*, Macmillan, London.

Motte, A. (1997), 'The institutional relations of plan-making', in P. Healey, A. Khakee, A. Motte, and B. Needham (eds.), *Making Strategic Spatial Plans*, UCL Press, London, pp. 231–54.

Northern Ireland Office (1999), *Shaping Our Future. Towards a Strategy for the Development of the Region*, Belfast.

Parry, R. (1992), 'The structure of the Scottish Office 1991', in L. Paterson and D. McCrone (eds.), *The Scottish Government Yearbook*, Edinburgh University Press, Edinburgh, pp. 247–55.

Parry, R. (1999), 'Quangos and the structure of the public sector in Scotland', *Scottish Affairs*, vol. 29, pp. 12–27.

Paterson, L. (1994), *The Autonomy of Modern Scotland*, Edinburgh University Press, Edinburgh.

Peat, J. and Boyle, S. (1999), *An Illustrated Guide to the Scottish Economy*, Duckworth, London.

Roberts, P.W. and Lloyd, M.G. (2000), 'Regional Development Agencies in England. New strategic regional planning issues?', *Regional Studies*, vol. 34 (1), pp. 75–81.

Rogers, S., Smith, M., Sullivan, H. and Clarke, M. (1999), *Community Planning in Scotland. An Evaluation of the Pathfinder Projects*, COSLA, Edinburgh.

Scottish Executive (2000a), *Social Justice. A Scotland Where Everyone Matters*, Edinburgh.

Scottish Executive (2000b), *The Way Forward: Framework for Economic Development in Scotland*, Edinburgh.

Sinclair, D. (1997), 'Local government and a Scottish Parliament', *Scottish Affairs*, vol. 19, pp. 14–21.

Skelcher, C. (1998), *The Appointed State. Quasi-governmental Organisations and Democracy*, Open University Press, Buckingham.

Town and Country Planning Association (1993), *Strategic Planning for Regional Development*, TCPA, London.

Turok, I. and Hopkins, N. (1998), 'Competition and area selection in Scotland's new urban policy', *Urban Studies*, vol. 35 (11), pp. 2021–61.

Wannop, U. (1986), 'Glasgow/Clydeside: a century of metropolitan evolution', in G. Gordon (ed.), *Regional Cities in the UK, 1890–1980*, Harper and Row, London, pp. 83–98.

Wannop, U. (1995), *The Regional Imperative*, Jessica Kingsley, London.

Wannop, U. and Cherry, G. (1994) 'The development of regional planning in the UK', *Planning Perspectives*, vol. 9, pp. 29–60.

# 5 Government Offices for the Regions and Regional Planning

MARK BAKER

## Introduction

A simple, but fundamental, question that can be asked of any system of regional planning is 'whose plan is it?'. In other words, where do the powers for preparing and implementing spatial planning at the regional level actually lie? In governmental systems that have a well-established (and usually elected) regional tier, the answer is often simple. Thus, within the federal structure of Germany, responsibility for regional planning falls to the *Länder*; Dutch Provinces prepare indicative spatial plans for their areas; Spanish regions (*Communidades Autonomas*) prepare their own regional plans, and so on.

The answer in countries that do not have regional government is, of course, much more complicated. In a highly centralized system, spatial planning at the regional scale might, for example, be undertaken and imposed by central government. In France, before the introduction of elected government at the regional level in the early 1980s, any such regional planning activity would normally have been carried out under the direction of the *Préfectures*, a powerful arm of central government in the regions since Napoleonic times. Alternatively, a bottom-up approach might be envisaged, with regional plans being jointly prepared by neighbouring local authorities: an example of this can also be drawn from current French arrangements where groups of Communes, often covering metropolitan areas, can collaborate to produce general planning objectives for their combined area (*Schéma Directeur*) (European Commission, 1994). A third approach might involve joint working between the centre and the local authorities within a region. It can be argued that the current approach to regional land-use planning in the UK falls into this third category but, in such a system, there is inevitably much scope for tension between the different tiers of government. Thus the issue of which tier is most influential, if central-local disagreement breaks out, becomes of crucial importance. This chapter aims to explore these issues by focusing on current regional planning processes in the English regions and, in particular, on the role of the Government Offices for the Regions (GORs).

## The Historical Context of English Regional Planning

Attempts at regional planning in the UK have, historically, followed different models because there was usually no obvious existing regional mechanism which could be utilised. The earliest examples, dating from the inter-war years, tended to involve voluntary associations of local authorities that oversaw the production of regional or sub-regional plans, with the work itself often undertaken by consultants (Wannop and Cherry, 1994). In contrast, Abercrombie's famous *Greater London Plan* (1944) was commissioned by central government and helped to influence the subsequent New Towns Act (1946) and Town and Country Planning Act (1947). During the 1960s and 1970s, examples can be found of regional plans and studies that were either statements of central policy (for example, the regional development plans for the North East and Central Scotland which came out as Government White Papers in 1963), special studies commissioned by central Government (such as the *South East Study* from 1964), or the products of local collaboration (such as the sub-regional studies which emerged in the late 1960s and early 1970s for neighouring counties such as Nottinghamshire and Derbyshire).

One common issue with regional strategy development, where there is no existing regional administration to undertake the task, is that there may be problems in terms of the subsequent ownership and implementation. From a bottom-up perspective, even if regional consensus across local authorities can be achieved, there may still be little chance of successful implementation if central government does not support the proposals and, where appropriate, make the necessary resources available. For example, the *Strategic Plan for the Northern Region* (1977) set out ambitious proposals involving economic policy and resource allocation which gained the support of regional partners, but ultimately lacked sufficient means of implementation without central government support. Such implementation gaps can also exist with a centrally-prepared strategy. As Hall (1992) reflects, the *Strategy for the South East* (1967), a product of the South East's Economic Planning Council (essentially a creation of central government), ran into opposition from the Standing Conference of Local Authorities within the region, concerned that their local autonomy in planning matters was being undermined. One obvious means of resolving such difficulties is to establish mechanisms where both central and local partners are fully involved in strategy formulation. Indeed, the next study that focused on the South East, the *Strategic Plan for the South East* (1970) was jointly commissioned by the Standing Conference and the Planning Council and was subsequently more successful.

The mechanisms for the preparation of English Regional Planning Guidance which have evolved since the late 1980s/early 1990s also have elements of central-local partnership in regional strategy production. The next section of this chapter examines the nature of this partnership in more depth and, particularly, focuses on the role of the GORs which are increasingly seen as having a partnership role with an ever-increasing number of other regional/local institutions and bodies but, nevertheless, also have a remit to implement national planning policy at the regional and local level.

**The English Regional Planning Guidance (RPG) Process**

English regional planning, in the form of RPG issued by the Secretary of State for the Environment, first emerged in the early 1990s. This adopted the model established for the preparation of strategic planning guidance (SPG) in London and the metropolitan areas following the abolition of the metropolitan county councils and the subsequent strategic gap in planning policy. The preparation mechanism involved an initial input from regional associations of local authorities (and, at least in some regions, additionally from other regional stakeholders through some form of regional conference arrangement), who were charged with submitting advice to the Secretary of State on the content of RPG for their region. From this point onwards, however, the preparation of RPG became primarily the role of central government, with the final RPG issued by the Secretary of State following consultation on an initial draft version. This process was criticised in several quarters for its over-centralised nature and its first outputs, such as the RPG for East Anglia (1991) and the North East (1993), were characterized as mainly bland re-statements of national planning policy that lacked any regional specificity (Minay, 1992). The process did, however, at least mark a renewal of interest by central government in regional policy development and helped to strengthen regional co-operation and partnership arrangements. There was also evidence of growing sophistication in both these regional conference arrangements and in the content of the subsequent RPG documents (Roberts, 1996; Baker, 1998).

More rapid change was, however, facilitated by the election in May 1997 of the new Labour administration which came into office committed to progressing newly devolved government structures in Scotland and Wales and new regional institutional arrangements in the English regions. As well as establishing Regional Development Agencies (RDAs) and encouraging the formation of non-elected Regional Chambers, proposals were also put forward for a revamping of the regional planning guidance process (DETR, 1998). These proposals were subsequently carried forward into a draft planning policy guidance note (DETR, 1999). Although the finalised version of PPG11 was not published until October 2000 (DETR,

2000), the RPG preparation process had essentially already been adopted by the Department of the Environment, Transport and the Regions (DETR) with the release of the draft Planning Policy Guidance (PPG) PPG11 and is therefore already reflected in the various arrangements which are currently being used in each of the English regions to revise their earlier RPG documents.

## The New Procedural Arrangements and the Role of GORs

GORs were created in April 1994 by the merger of the existing regional offices of four government departments – Environment, Transport, Trade and Industry, and Employment. There are currently nine GORs (Figure 5.1) and, in general, their administrative boundaries correspond with the regional units currently used for regional planning purposes (the main discrepancy being in the Eastern Region, where part of the area covered by the Government Office (GO) is included within the area covered by regional planning guidance for the South East and the remainder is subject to a separate RPG for East Anglia). The GOR boundaries also correspond with those of the newly established RDAs. The overall 'mission' of the GORs is: '... to work with regional partners and local people to maximize competitiveness and prosperity in the regions, and to support integrated policies for an inclusive society.' (Cabinet Office Performance and Innovation Unit 2000: Annex B.)

To help achieve this rather wide-ranging mission, they have responsibility for managing government programmes on behalf of parent Departments; identifying and facilitating effective linkages between partners and programmes; and informing the development of government policies from a regional perspective (Cabinet Office Performance and Innovation Unit (PIU, 2000)).

These statements, and the GORs' more detailed objectives, tend to stress a partnership approach with regional stakeholders. However, in practice, the role also extends beyond such joint working to take on board the implementation of central policy and, in some areas, regulation of local activities. This is certainly true in terms of the operation of the statutory planning system, where GORs are responsible for scrutinising emerging development plan policies at the local level to ensure that they comply with national planning policy. The Secretary of State has reserve powers to enforce compliance (Ho, 1997; Baker, 1999). Similarly, there are powers to 'call in' planning applications that would normally be determined by local planning authorities. The role of the GORs is, therefore, a somewhat complex one in which they are sometimes seen as just one of a number of regional partners working toward a common regional agenda, but on other occasions may exert their powers and authority to ensure other regional partners comply with national policy. These different aspects of GOR behaviour are also potentially inherent in the current regional planning arrangements.

One of the main stated purposes of the revised RPG arrangements, as set out in the introduction to PPG11 (DETR, 2000), is to place '...greater responsibility on regional planning bodies, working with the Government Offices and regional stakeholders, to resolve planning issues at the regional level through the production of draft RPG. This promotes greater local ownership of regional policies and increased commitment to their implementation through the statutory planning process ...'.

Other statements within the PPG also emphasize that, compared to the previous arrangements, regional planning bodies (essentially the existing regional local authority associations or conferences, but potentially newly-constituted regional chambers) have enhanced responsibilities (DETR, 2000, 1.5) and that it is essential that other regional stakeholders (including the new RDAs) should also be involved (DETR, 2000, 2.7). All this suggests that the revised RPG strategies will primarily belong to the region. Ultimately, however, the responsibility for issuing the final document still lies with the Secretary of State.

Throughout the RPG preparation process, the GORs (acting on behalf of the Secretary of State) are also actively involved. PPG11 (DETR, 2000, 2.6) provides guidance on this: 'The Government Offices will play an important role in the production of RPG ... including by advising on relevant national policy, proposing issues that need to be addressed, commissioning joint research and assisting in the collection of data.' This advice goes on to stress that GO involvement should be conducted in an '...open and transparent manner, with all advice being made public'.

The roles of the Government Office / Secretary of State are set out in more detail in which is sourced from a similar table in PPG11 (DETR, 2000, Table 5.1). This reveals that the Secretary of State / GO is actively involved in all the preparation stages although there is a significant switch in the nature of the working relationships between the GO and the regional planning body (RPB) from around stage four (public examination) onwards. Up to this point, the table highlights various forms of 'co-operation' and 'consultation' between the GO and the RPB. From this point onwards, however, central government takes primary responsibility for progressing the RPG and the table thus details the more formal role of the Secretary of State in proposing changes to the draft RPG and issuing the final RPG. Once issued, the GOR also has a role in ensuring that development plans and transport plans at the local level accord with the regional strategy. The GO also liaises with the RPB, and other regional partners, over appropriate monitoring and review arrangements.

**Figure 5.1: Map of Government Offices for the Regions**

*Source:* Cabinet Office Performance and Innovation Unit, 2000

In a procedural sense, therefore, the respective roles of the RPBs, GORs and the Secretary of State are fairly clear. Nevertheless, there still remains some scope for variation in the ways these roles are exercised. In this respect, PPG11 gives few clues as to how the relationships between the GOs and RPBs operate in practice; nor does it adequately explain, if there are any apparent central-local differences in policy views, how such tensions are to be resolved (and which side usually wins!). In an attempt to get behind such issues, this chapter now looks at the on-going RPG preparation process in the North West region which provides a good example of the role of a GO in the early stages of RPG preparation when primary responsibility for progressing the strategy still lies with the RPB. This is followed by some thoughts on the role of the Secretary of State/GORs in the latter stages of RPG preparation following the public examination.

Two characteristics mark out the RPG preparation process in the North West as somewhat different from that in most other regions. Firstly, a team was established with responsibility for undertaking the technical work associated with RPG preparation. This Regional Planning Guidance team is based in the Policy Unit offices of the North West Regional Assembly (NWRA) in Wigan and consists of a full-time Regional Policy Co-ordinator with responsibility for around six or seven other professional staff. Thus, unlike several other regions where responsibility for RPG work has generally fallen to working groups consisting of planning officers who also continued their normal 'day-jobs', the work in the North West has been able to benefit from a dedicated, single-minded, and perhaps more independent, team of officers. The second feature of the early phases of RPG work in the North West has been the commissioning of a series of research studies which were intended to provide a direct input to RPG. Although other regions have also initiated some regionally-based research, the North West examples are especially notable for their extent and range of topics covered (Table 5.2).

**Table 5.1: Key Stages in Regional Planning Guidance (RPG) Preparation, Showing Role of Government Office/Secretary of State**

| | |
|---|---|
| **Overall summary of process:** | RPG prepared by regional planning body (RPB), *in co-operation with Government Office (GO)*, and *issued by the Secretary of State (SoS)*. |
| *Stage 1: identifying the brief / issues* | RPB *in consultation with GO* draw up the project brief ... and hold a one-day public conference to seek agreement. |
| *Stage 2: developing the draft strategy* | RPB *in co-operation with GO* and in consultation with regional stakeholders carries out technical/survey work.... and then develops and refines options into draft RPG. |
| *Stage 3: consultation on draft RPG* | RPB *submits draft RPG and appraisal to SoS*. Draft strategy published for consultation. Written responses (to Panel Secretary) copied to *GO*/RPB. |
| *Stage 4: testing (public examination)* | Panel *in consultation with RPG and GO* agree list of matters and invite participants to public examination. Draft RPG tested by independent Panel *appointed by SoS*. |
| *Stage 5: publication of Panel report* | *Panel reports to SoS* (which GO copies to RPB and other public examination participants). |
| *Stage 6: proposed changes to RPG* | *SoS (GO) publishes proposed changes to draft RPG*, with a statement of reasons, for consultation. |
| *Stage 7: issue of final RPG* | *RPG approved and issued by SoS.* |
| *Stage 8: development / local transport plan conformity* | SoS (GO), with support of RPB, *to ensure development and local transport plans are consistent with RPG.* |
| *Stage 9: monitoring and review* | RPB, *in liaison with GO*, establish monitoring machinery to check on achievement of RPG targets and review it as appropriate. |

*Source:* Adapted from DETR, 2000, 2.13

**Table 5.2: North West Regional Planning Guidance: Commissioned Research Studies**

| Research Study | Consultants | Main partners |
| --- | --- | --- |
| *North West Housing Need & Demand Research* (final report published by DETR, March 2000) | Manchester & Liverpool Universities | GONW/DETR (f, s)<br>NWRA (f, s)<br>National Housing Federation (f, s)<br>Northern Consortium of Housing Authorities (f, s)<br>House Builders Federation (f, s)<br>Countryside Agency (f, s)<br>Halton Borough Council (f, s)<br>Housing Corporation (f, s) |
| *Research into Integrated Coastal planning in the North West Region* (final report published by DETR, May 2000) | Liverpool University | GONW/DETR (f, s)<br>NWRA (s)<br>English Nature (s)<br>Environment Agency (s)<br>PISCES (s)<br>Barrow Borough Council (s) |
| *England's North West: a Strategy Towards 2020 and RPG Review – Spatial Implications of the North West Regional Strategy* (final report published by NWDA / DETR, 2000) | DTZ Pieda | GONW/DETR (f, s)<br>NWRA (f, s)<br>NWDA (f, s) |
| *Research into North West Regional Planning Guidance Car Parking Standards* (unpublished final report, March 2000) | Oscar Faber | GONW/DETR (f, s)<br>NWRA (f, s) |
| *A Report on Urban Fringe which is Green Belt Land in the North West* (unpublished final report, March 2000) | Baker Associates and Enderby Associates | GONW/DETR (f, s)<br>NWRA (f, s)<br>CPRE (s)<br>NWDA (s) |
| *Beyond the Millenium: Debate of the Age Demographic research* (final report published by NWRA, 2000) | Lancaster University | GONW/DETR (f, s)<br>NWRA (f, s) |

Notes: f = funding input; s = steering group membership
Source: Various research reports and personal communication with GONW

Although the NWRA has obviously had a considerable involvement in both establishing the RPG Team and commissioning research, there is also little doubt that the full realisation of these matters has been possible because of the encouragement, and financial support, given by DETR/Government Office for the North West (GONW). The significance of this close working relationship between the NWRA and the GONW is explicitly recognised by NWRA in the foreword to the first major consultation document to emerge from the RPG revision process (NWRA, 1999a) which states that, '... the Assembly has worked closely with the Government Office for the North West and other regional stakeholders to put into place a professional team to develop this crucial strategy. In this the North West is taking the lead nationally ...'. Thus part of the financial package behind the RPG team has come from central government sources (including, for example, the salary costs for a seconded local government officer). Equally, the DETR/GONW has also made a major contribution towards the research studies which have provided an input to the RPG process. Indeed, several of these studies were fully funded by the DETR/GONW and, even where joint funding was involved, the DETR/GONW typically contributed the largest amount. Of the research studies that have been published, two were done so through the DETR's in-house research publication series (Wong and Madden, 2000; Kidd et al, 2000).

The process of RPG preparation in the North West region has broadly followed that set out in PPG11 (DETR, 2000). The main outputs in this process have been consultation exercises in April 1999 on the issues to be covered in the revised RPG (NWRA, 1999a); in October 1999 on the strategic options for the region (NWRA, 1999b) and finally, in June 2000 the publication of draft RPG for the North West NWRA, 2000). Compared with some other regions, the consultation process in the North West was significant in terms of its extent (in terms of the numbers of consultees/respondents) and length of time taken – indeed the original timetable for the preparation process was extended to allow sufficient consultation time for the document 'Choices for the North West'. Responsibility for preparing these documents, and the associated consultation processes and responses, lay mainly with the RPG team in Wigan. However, at the political level, the work of the RPG team was overseen by NWRA's Planning, Environment and Transport Key Priority Group which consisted of elected local authority members and some non-local authority representatives, and fed back into the NWRA's Policy Committee. At officer level, there is an RPG Officers Steering Group that includes planning officers representing local authorities across the region, supplemented by representatives of other regional interests such as the NWDA. Helping the RPG team in their work, a number of Task Groups were also created which involved representatives from a number of regional interests and covered specific topics: housing and population; transport; environment; and economy (Jenkinson, 1999). Steering Groups of various sizes were also established to oversee the individual research studies.

During this process, GONW were represented on the RPG Officers Steering Group and all of the Task Groups and research steering groups. They also had observer status on NWRA's Key Priorities Groups. In addition, the RPG team also organised regular liaison sessions with key regional agencies and organisations, which included GONW. In order to provide input to the RPG process, a small team of 3–4 civil servants within the Planning and Environment Group, based in GONW's offices in Sunley Tower, Manchester, was responsible for RPG related work, including both liaison with the RPG team in Wigan and the administration of the RPG research programme. Thus, throughout the RPG preparation stages up to, and including, consultation on the draft RPG, the role and input of GONW can be split into three categories or levels:

- 'official' responses to the consultation documents during the stated consultation periods;
- influence through formal membership of the RPG Officers Steering Group, the four Task Groups and the various research steering groups (in the case of the latter, often as Chair and the main client and/or contract manager);
- other *ad hoc*, informal input through on-going, and frequent, liaison with the RPG team.

The first level of input is clearly open to all interested parties. The second is also open to a relatively large number of stakeholders/regional interests as representation via the Task Groups does appear to cover a broad cross-section of organisations and regional interests. Membership of the Officer Steering Group and the individual research steering groups was, however, more tightly drawn with NWRA, NWDA and GONW emerging as the key players, and represented on most groups (but not quite all in the case of NWDA). It is therefore through the third level of more informal liaison that GONW can perhaps exert special influence and, although direct liaison was carried out by the RPG team with a number of regional stakeholders (for example NWDA, HBF, CPRE and so on), the linkages between the respective RPG teams in Wigan and Manchester have been particularly strong.

So, throughout the early stages of the RPG process, and before DETR/GONW takes on direct responsibility for its preparation post public examination, a variety of mechanisms exist that allow GONW to influence the emerging draft RPG. This has extended well beyond the normal consultation periods; for example, GONW have been in a position to comment, and potentially influence the content of, a large number of working drafts of documents before they were officially published. It should also be remembered that, since responsibility for issuing the final RPG actually rests with the Secretary of State, there may be a tendency on behalf of regional planning bodies to take more account of GOR comments on emerging drafts than other, less powerful, stakeholders. Indeed, this special position of the GOR appears to have been recognised by some other stakeholders who sometimes lobbied GONW directly in respect of the emerging RPG as well as providing their responses to NWRA via the 'official' consultation channels. Nevertheless, despite the close linkages between the two RPG teams and the special status of DETR/GONW, it should not be assumed that NWRA has necessarily taken on board all of GONW's interests and views on the content of the draft RPG. Indeed, at times, tensions between the RPG team and GONW appear to have arisen. One such area of disagreement occurred over the presentation of the strategic options to be included in the Choices for the North West consultation paper (NWRA, 1999b) although, on this occasion, it does appear that direct liaison by GONW with the RPG team did help resolve some of the outstanding issues prior to the paper's publication.

Draft revised RPG for the North West (NWRA, 2000) was finally submitted to the Secretary of State and simultaneously published for consultation purposes in July 2000. In their submission letter, NWRA make specific mention of the close links with GONW during these early preparation stages: '... At all stages, we have liaised closely with colleagues in GONW and would like to thank them for their valuable assistance.' Nevertheless, the content of the document remains primarily that of NWRA and, as can be seen by GONW's response to the consultation (GONW, 2000) there are a number of outstanding issues to be resolved. In this response, GONW not only highlight a number of relatively important general issues (Table 5.3) but also include a detailed schedule of comments which stretches to over 60 pages! Thus, despite the extent of two-way liaison over its preparation, there are still a large number of areas where the content of the draft RPG is still considered by GONW to either conflict with national policies or contain significant omissions. The next step is for the more fundamental of these differences, along with issues highlighted by other respondents, to be tested at the public examination, in February 2001. A Panel Secretariat was established by GONW to administer the responses to the draft and prepare for the public examination. Perhaps as a result of the extent of consultation undertaken at the earlier stages, it is having to deal with over 400 responses – a high figure for a regional strategy document and significantly more than in most other regions including neighbouring Yorkshire and Humberside. Following the public examination, GONW/Secretary of State will then take over primary responsibility for progressing the revised RPG, with the final RPG expected to be issued by the Secretary of State by the end of 2001.

**The Role of GORs in the Latter Stages of RPG Preparation**

Whereas the North West provides an ideal example to illustrate the role of a GOR in the early stages of RPG preparation, because it has not yet reached the public examination, it is necessary to look to other regions for examples of GOR/Secretary of State involvement in the latter stages of preparation. At the time of writing, there are still no regions where the final RPG has been issued by the Secretary of State. However, the preparation process in East Anglia and the South East is well-advanced and it is thus possible to draw some pointers from a fairly brief examination of recent events in these two areas.

The revised RPG for East Anglia had its origins in a Regional Strategy for East Anglia (SCEALA, 1997) which was submitted as advice to the Secretary of State for the Environment on the content of RPG by the Standing Conference of East Anglian Local Authorities (SCEALA) in August 1997. However, in the light of the new procedures which were emerging from the new Labour government at around this time, it was agreed that SCEALA's earlier strategy should be re-issued for consultation purposes as draft RPG in August 1998 (SCEALA, 1998). The consultation process was organised by the Government Office for the Eastern region (GO-East). Responsibility for the selection of issues and participants for the subsequent public examination fell primarily to the Panel appointed by central government to conduct the examination, but also included input from both GO-East and SCEALA. The Panel's Report was submitted to the Secretary of State/GO-East in July 1999.

In March 2000, the Secretary of State published, for consultation purposes, his proposed changes to the draft RPG (GO-East, 2000). These were set out in a lengthy schedule as well as incorporated into a re-printed version of the full RPG. Although many of these changes reflected recommendations made by the Panel who conducted the public examination, there were also a number of areas where the Secretary of State either rejected the Panel's recommendations (and proposed different changes) or proposed new changes in the light of national government policy. The accompanying letter from the Planning Minister, Nick Raynsford, to SCEALA indicated that he would be particularly interested in the views of SCEALA, as author of the draft RPG, on the proposed changes. However, this letter also suggested that SCEALA might be surprised at the extent of the changes; '... the changes are perhaps greater than you might have anticipated'. Significant changes included slightly lower housing requirements than those initially suggested by SCEALA (and significantly lower than the higher figures put forward by the Panel) and a higher target for residential development of brownfield land of 50%. This contrasted with SCEALA's earlier figure of 40% which had also gained the support of the Panel. The proposed changes thus demonstrated the Government's willingness to make a significant number of alterations, in many cases of a detailed nature, and that they were prepared to do so even if the Panel were not in agreement. Nevertheless, despite reservations over whether the higher brownfield targets would be achievable, the initial reaction of SCEALA to the proposed changes appeared to be fairly positive (Harfield, 2000).

**Table 5.3: Main Issues Raised in GONW's Response to Draft RPG for the North West Region**

- *Structure of the RPG document*
- *Regional transport strategy and transport accessibility*
- *The European context and the European Spatial Development Perspective*
- *Core strategy*
- *The strategic framework*
- *Sub-regions*
- *Treatment of Regional Investment Sites*
- *Tourism, sport and leisure and cultural heritage interest*
- *Housing provision*
- *Green belt*
- *Coastal vision and need for a chapter on coastal matters*
- *Climate change*
- *Targets and indicators*

*Source:* Adapted from GONW, 2000

In contrast, the process in the South East has been considerably more controversial. In this case, the London and South East Regional Planning Conference (SERPLAN) submitted their 'Sustainable Development Strategy for the South East' (SERPLAN, 1998) to the Secretary of State in December 1998. This was treated as draft RPG, and the Government Office for the South East (GOSE) organised a consultation process which ran until February 1999. By this time, the Secretary of State had also appointed the

Panel which was to conduct the subsequent public examination. Reflecting the special circumstances of the South East, there are three separate GORs with interests in the RPG for the South East – GOSE, Government Office for London (GOL), and GO-East (which incorporates Essex, Bedfordshire and Hertfordshire as well as East Anglia which is covered by the separate RPG which has already been described above). Along with SERPLAN, all three Government Offices provided their views on the selection of participants and issues for the public examination, which were formally determined by the Panel.

The public examination was held in May 1999 and the GORs were invited participants at all eleven of the broad topic areas chosen for discussion. The Panel's Report was submitted to the Secretary of State in September 1999 and, when published in October 1999 (Crow and Whittaker, 1999), raised considerable debate and controversy over its recommendations, particularly in respect of overall housing requirements and the extent to which the South East should be allowed to fulfil its economic potential through further development. In doing so, it also set out significant criticisms of SERPLAN's draft RPG.

Whilst the Secretary of State and relevant GOs considered their response, arguments raged in both the local media and professional journals over the merits, or otherwise, of the Panel's recommendations in what became known as the 'Crow Report' after the Panel Chair, Professor Stephen Crow (for example see Garlick, 2000; Hall, 1999). Eventually, on 7 March 2000, the Deputy Prime Minister (who is also the Secretary of State) made a statement to the House of Commons covering the Government's intentions in respect of both revised national planning guidance for housing, and the revised RPG for the South East. This criticised both SERPLAN – for failing to take account of future housing needs or account for the growth of single person households – and the Crow Panel – for applying a rigid 'predict and provide' approach to housing and for not paying enough attention to the capacity of London and the South East to absorb and plan for growth sustainably (Prescott, 2000). Instead, the Government's statement called for housing provision figures of around 43,000 per annum. This was higher than SERPLAN's suggestion in draft RPG of around 33,000 dwellings per year but significantly less than the Panel recommendation of some 55,000 dwellings per year. Although the debates over housing requirements have dominated the subsequent discussions of the Government's proposed changes to SERPLAN's draft RPG, there were also a substantial number of other changes that were set out in consultation documents published by the three Government Offices in March 2000 (GOSE *et al*, 2000). As with the RPG for East Anglia, some of these reflected the Panel's recommendations, but others were introduced by the Secretary of State/GORs without the Panel's support.

## Conclusions

This chapter has examined the role of the GORs/Secretary of State in the new arrangements for preparing regional planning guidance. The previous arrangements for RPG preparation were widely criticised for their degree of centralisation – with the outputs often considered little more than a rehash of national planning policies. The new arrangements, introduced by the newly-elected Labour administration, appear to place more emphasis on regional ownership of the emerging RPG strategies and incorporate an element of independent scrutiny via a public examination. Nevertheless, a more detailed examination of the different steps in the preparation process reveals that central government (GOR/Secretary of State) still has a significant, and decisive, role to play throughout the process.

Even in the early stages up to the public examination, where primary responsibility for RPG work lies with the regional planning body, the North West example shows how the GOR can have considerable influence via a number of routes including financial support for research studies, representation on key decision-making/advisory bodies, and regular liaison outside 'official' consultation periods. Their special status, as ultimate arbiter of what is included in the final RPG document, may also encourage regional planning bodies to take their views particularly seriously. The examples of the two regions which have progressed beyond the public examination to the point where the Secretary of State has now published his proposed changes reveals the ability, and willingness, of the centre to make numerous, and significant, changes to draft RPGs, even if they are not always backed up by supporting Panel recommendations.

Whatever the perceived merits of SERPLAN's draft RPG (SERPLAN, 1998) and the subsequent 'Crow Report' (Crow and Whittaker, 1999), the Government's response (and indeed its manner via a Deputy Prime Minister statement to the House of Commons) illustrates that the eventual outcome of controversial issues at the regional level remains largely a political decision by the centre. In the South East, the tensions between the views of central government and those of SERPLAN/local authorities on the content of RPG, and particularly the housing requirements, appear likely to continue up to and beyond the issuing of the final RPG (Dewer, 2000), and may well re-surface at the local/sub-regional level if individual structure plans and UDPs fail to deliver the required number of dwellings. In East Anglia, however, there appears to be a greater level of agreement and thus a greater chance of local support and implementation. With the preparation processes in other regions still at various stages, it remains too early to determine the extent to which the key stakeholders in each region will support, and claim ownership, of the final RPG strategies. Nevertheless, whilst it is probably fair to describe the new RPG arrangements as a joint partnership between regional stakeholders and central government (rather than predominantly 'top-down' or 'bottom-up'), the experiences to date suggest that, if any party does end up unhappy with the eventual outcome, it is unlikely to be the GORs/Secretary of State!

## References

Baker Associates and Enderby Associates (2000), *A Report on Urban Fringe which is Green Belt Land in the North West*, unpublished RPG research report – for details contact GONW, Sunley Tower, Manchester, M1 4BE.

Baker, M. (1998), 'Planning for the English regions: a review of the Secretary of State's regional planning guidance', *Planning Practice and Research*, vol. 23, pp. 153–69.

Baker, M. (1999), 'Intervention or interference? Central government involvement in the plan-making process in six English regions', *Town Planning Review*, vol. 70 (1), pp. 1–24.

Cabinet Office Performance and Innovation Unit (2000), *Reaching Out*, HMSO, London.

Crow, S. and Whittaker, R. (1999), *Regional Planning Guidance for the South East of England: Public Examination May-June 1999 – Report of the Panel*, GOSE, Guildford.

Department of the Environment (1996), *RPG13: Regional Planning Guidance for the North West*, HMSO, London.

Department of the Environment, Transport and the Regions (1998), *The Future of Regional Planning Guidance*, DETR, London.

Department of the Environment, Transport and the Regions (1999), *Planning Policy Guidance Note 11: Regional Planning (Public Consultation Draft)*, DETR, London.

Department of the Environment, Transport and the Regions (2000), *Planning Policy Guidance Note 11: Regional Planning*, DETR, London.

Dewer, D. (2000), 'Conflict brews over new housing figure', *Planning*, vol. 1372, 9 June 2000, p. 1.

DTZ/Pieda, (2000), *Spatial Implications of the North West Regional Strategy*, NWDA, Warrington.

European Commission (1994), *Europe 2000+: Cooperation for European Territorial Development*, European Commission, Luxembourg.

Garlick, R. (2000), 'DETR told to justify extra housing total', *Planning*, vol. 1363, 7 April 2000, p. 3.

Government Office for the Eastern Region (2000), *Draft RPG for East Anglia: Proposed Changes by the Secretary of State for the Environment, Transport and the Regions*, GO-East, Bedford.

Government Office for the North West (2000), *Public Examination into Draft Review of RPG for the North West: Comments from the Government Office for the North West*, GONW, Manchester.

Government Offices for the South East, London and Eastern Regions (2000), *Draft Regional Planning Guidance for the South East (RPG9): Proposed Changes*, GOSE, Guildford.

Hall, P. (1992), *Urban and Regional Planning* (3$^{rd}$ ed.), Routledge, London.

Hall, P. (1999), 'Braving a call to war – and adding to John Prescott's headaches', *Town and Country Planning*, vol. 68 (12), pp. 356–7.

Harfield, R. (2000), 'City's green belt at risk in regional plan', *Planning*, vol. 1362, 31 March 2000, p. 3.

Ho, S.Y. (1997), 'Scrutiny and direction: implications of government intervention in the development plan process in England', *Urban Studies*, vol. 34, pp. 1259–74.

Jenkinson, I. (1999), *The Institutional Basis for Regional Strategic Planning: the Contribution of Regional Planning Guidance for the North West*, unpublished Masters dissertation, School of Planning & Landscape, University of Manchester.

Kidd, S., Massey, D., Hough, A. and Cousins, N. (2000), *Research into Integrated Coastal Planning in the North West Region*, DETR, London.

Lancaster University (2000), *Beyond the Millennium: Debate of the Age Demographic Research*, NWRA, Wigan.

Minay, C. L. W. (1992), 'Developing regional planning guidance in England and Wales: a review symposium', *Town Planning Review*, vol. 63, pp. 415–34.

North West Regional Assembly (1999a), *Regional Planning Guidance: Listening to the North West*, NWRA, Warrington.

North West Regional Assembly (1999b), *Regional Planning Guidance Review: Choices for the North West*, NWRA, Warrington.

North West Regional Assembly (2000), *People, Places and Prosperity: Draft RPG for the North West*, NWRA, Warrington.

Oscar Faber (2000), *Research into North West Regional Planning Guidance Car Parking Standards*, unpublished RPG research report – for details contact GONW, Sunley Tower, Manchester, M1 4BE.

Prescott, J. (2000), *PPG3 and Planning for the South East*, Deputy Prime Minister's Statement to the House of Commons, Tuesday 2 March 2000.

Roberts, P. (1996), 'Regional planning guidance in England and Wales: back to the future?', *Town Planning Review*, vol. 67(1), pp. 97–109.

SERPLAN (London and South East Regional Planning Conference) (1998), *A Sustainable Development Strategy for the South East*, SERPLAN, London.

Standing Conference of East Anglian Local Authorities (SCEALA) (1997), *Regional Strategy for East Anglia*, SCEALA, Norwich.

Standing Conference of East Anglian Local Authorities (SCEALA) (1998), *Draft Regional Planning Guidance for East Anglia*, SCEALA, Norwich.

Wannop, U. and Cherry, G. (1994), 'The development of regional planning in the United Kingdom', *Planning Perspectives*, vol. 9, pp. 29–60.

Wong, C. and Madden, M. (2000), *North West Regional Housing Need and Demand Research*, DETR, London.

# 6 Integrating Local and Central Concerns in the New Regional Landscape: Pipe Dream or Political Project?

MARTIN STOTT

**Introduction**

National and regional questions have been a long-established feature of UK political life. Irish 'Home Rule' dominated Victorian political life, and more recently the failure of the Callaghan Government in the 1970s to fully recognise the implications of their proposed devolution to Scotland, on the North East and other regions led, eventually, to the wrecking of the Government's Devolution Bill and ultimately the fall of the Government in 1979.

Since then, two factors have changed the character of the 'regional debate' in the UK. Firstly, the creation by the new Labour Government of an elected parliament/assemblies in Scotland, Wales, Northern Ireland and London, and secondly, pressure from the European Union (EU) as France, Italy and Spain have joined Germany in creating a significant regional government structure – 'Europe of the regions'. The then Government's response to this pressure was the creation of Government Offices for the Regions (GORs) in 1994. These domestic and international reforms along with a more general reappraisal of what kind of functions are most appropriately performed at national and local levels in an era of globalisation have given rise to a debate over what is sometimes referred to as 'the English question'.

This chapter will examine not so much what has been referred to as 'the missing middle' (New Local Government Network, 2000) in the debate around the English question, but related questions of consistency and 'joined upness' in government policy, and the implications this has for effective and accountable government at the local level, focussing particularly on the South East and Thames Valley.

## Modernising the Institutions

The 'silo mentality' or departmentalism in Whitehall has been a long-running criticism both by academic commentators and by many in local government who are concerned with the territorial management of public policy. The establishment of GORs was a first attempt to do something about this. However, their very establishment saw the contradictions in Government thinking exposed. The regions themselves were arbitrary. The SERPLAN South East was split into three, London, Eastern and South East, with the result that major regeneration programmes such as the Thames Gateway were split between three GORs. Their size hardly assisted the creation of a common identity – the South West is so large that parts of it are closer to Scotland than they are to other parts of the same region. Furthermore, the attitude of Whitehall was at best ambivalent as some departments – Department of Trade and Industry (DTI), Department of Environment Transport and the Regions (DETR), Department for Education and Employment (DfEE), participated while others, for example the Ministry of Agriculture Fisheries and Food (MAFF), and the Department of Health (DoH) did not, and indeed the DoH maintained a regional structure based on different geographical units.

Although a senior civil servant was appointed to head up each GOR, their budgets none the less remained firmly in the control of each sponsoring department, to which separate bids have to be made annually by each GOR – more Whitehall in the regions than the regions in Whitehall.

The change of Government in 1997 sparked a real concern about how, in the process of modernising the institutions of the state, 'joined up government' at the local and regional level could be made a reality. The Government's 'Modernising Government' (Cabinet Office, 1999) programme recognises a need for more joined up government at the regional level. A central recommendation is that all public bodies should work towards common boundaries based on those of local government and the GORs. More recently the Performance and Innovation Unit report (Cabinet Office Performance and Innovation Unit, 2000) has identified problems of multiple conflicting government initiatives at local level.

The Government's reforms of sub-national institutions have variously strengthened, simplified and democratised the sub-national administrations in the non-English nations and in London (where the model is more about local government reform – the return of a post GLC London wide authority – than devolution). However, in the rest of England, the reforms so far, with a heavy centralising steer, and a tendency to remove local services from local accountability, have added to the complexity of sub-national governance. The kinds of issues that are best dealt with regionally or locally in England include

- aspects of economic performance, particularly skills;
- social inclusion;
- the imbalance between housing needs and costs, and associated long distance commuting;
- congestion and investment in transport infrastructure;
- health inequalities;
- environmental and quality of life issues including emerging regional cultural strategies, waste management, tackling the trends towards de-urbanisation in both housing and jobs; and
- the related issue of whether an 'urban renaissance' is possible and how to deliver it.

In many of these areas the gap between government words and government actions has been striking. The stated push towards 'joined up government' is coming just at the time when at the local level it is being taken apart as more and more former local government functions are being hived off to quangos (Stott, 2000a). The DfEE are particularly 'off message' in this respect. As well as having an Employment Service whose local areas bear no resemblance either to regional structures or local government jurisdictions, they are in the process of dismantling Training and Enterprise Councils and replacing them with Local Learning and Skills Councils based on areas such as Berkshire (now six unitary authorities) and Milton Keynes / Buckinghamshire / Oxfordshire (three local education authorities). These are not coterminous with either their Regional Development Agency (RDA) and regional assembly areas, or their principal local authorities. This same process has been mirrored by the DTI with the creation of the new Small Business Service (SBS) where franchising decisions have ignored not only local authority jurisdictions but RDA ones as well, for example the award of the Berkshire franchise (South East) to the Swindon and Wiltshire SBS (South West). This is despite the appeals of the RDAs and even GOR civil servants, and has been happening all over the country. The RDA structure, however, is seen as a creature of the DETR – a rival department in Whitehall, and local authorities are seen as part of the problem rather than part of the solution.

This view of local authorities is exacerbated by the tendency of Government departments to 'parachute in' all kinds of local initiatives. At the last count there were over 50 different government departments or agencies alongside over 30 'action zones' and other area initiatives established since the Labour Government came into office (Mawson, 2000). As well as sometimes unhelpful challenging of local policy priorities or targeting of similar client groups or areas, this leads to much duplication of effort. So much for joined up government at the local level. It is this continuing silo mentality in Whitehall that is the major source of the difficulties facing the successful establishment of coherent effective and acceptable sub-national governmental structures in England.

**City Regions: Regional Government with Attitude?**

In policy terms the waters have been further muddied by the emergence of an alternative to the RDA/Assembly-based regional agenda. Whilst one arm of the DETR has been promoting the RDA/Assembly model, those involved in local government modernisation have been promoting another – the City Region, particularly through the notion of powerful elected city mayors.

The Local Government Association (LGA) report (2000) makes the point that public acceptance of regions is a key component in the success of changes to present regional arrangements, and that without democratic legitimacy they are bound to fail. As the Government has always insisted that a positive result in a regional referendum is a necessary precondition before further moves towards elected regional government can take place, it is hardly a surprising conclusion. However, it raises important questions about the long-term stability of the present RDA/Assembly arrangements. The New Local Government Network (NLGN) report (2000) summarises the point quite succinctly stating that 'urban regions are the most obvious alternative to [RDA] regions in terms of geographical scale, because they represent more coherent economic entities and command greater affinity amongst their inhabitants'.

Interestingly the conference at which this report was launched was addressed both by the LGA Chief Executive and the Local Government and Regions minister Hilary Armstrong. In endorsing the report she argued that the concept of City Regions (where the name of the city gives national identity, and by implication local identity too) fitted well with, for example, the Government's commitment to urban renaissance. This is in contrast to views expressed about the likelihood of RDA/Assembly regions achieving such local identity and support.

It is clear that a rethink is underway in Government about the logical conclusions of the up to now promoted 'RDA/Assembly' model, leading to eventually elected regional government. The model has been the cornerstone of DETR thinking, itself heavily influenced by the report 'Renewing the regions' (Regional Policy Commission, 1996) commissioned by the Labour Party when in opposition. Instead, late in the day, a recognition is dawning that local government modernisation, democratic renewal and a viable regional government structure are linked. The question of how to square the trajectory of a revitalised local government, headed in many areas by elected city mayors with ambitions for their city regions, and elected regional assemblies is only now being considered. The idea that a city mayor in, for example, Birmingham or Newcastle could work easily with the leader of an elected regional assembly for the West Midlands or the North East, based respectively in Birmingham and Newcastle, is improbable (Stott, 2000b).

The question is where is the rethink going? What does it mean for the many parts of the country, including the whole of the South East outside London, and the South West outside the greater Bristol areas, which do not have a large city focus? In raising this issue one begs the question of the appropriate 'level' of responsibility for the large strategic functions such as regional planning and transport. Some city regions may be large enough to address these issues, others will not be.

The LGA has come up with six possible scenarios, and the NLGN with four. Neither organisation has a 'preferred model' though the trend is clearly towards a more pluralist model than the 'one size fits all' approach of the GOR/RDA/Assembly regions. Variations on a 'city regions, and where there is no "suitable" city, sub regions' model are clearly ones that find favour. The LGA report goes so far as to say '. . . the sub region emerges as a crucial unit, given its ability to bridge boundaries and to give a meaningful economic unit that can involve areas otherwise lying on the periphery of a region'.

This description fits exactly the position of a city like Oxford, clearly in the South East and therefore in the London city region – but probably not best served by being governed as part of London. The interesting question that will arise if this model is taken forward is, would the 'sub region' be 'Oxfordshire', the city region of Oxford, or the 'Thames Valley', whose 'capital' arguably is Reading and for which there is already in place a (non-statutory) sub-regional institution, the Association of Councils of the Thames Valley Region.

## Making Sense for Local Communities

The point at which existing local government jurisdictions merge almost imperceptibly into 'sub-regions' will be a tricky one for the LGA and others to manage.

The LGA report sets out the following four principles that should underline future regional arrangements:

- a greater devolution of powers in England based on a consolidation of subsidiarity;
- flexibility – not everyone has to go at the pace of the slowest and geographical variations are acceptable;
- transparency – more public scrutiny, fewer quangos;
- ensuring that future developments build on the strengths of the current developments – some progress so far has been positive, so don't throw the baby out with the bath water.

It also poses six questions, listed below, that it feels are necessary to assess whether future regional arrangements make sense for governing local communities:

- Would the proposals bring decision-making close to the people affected?
- Do the proposals bring decision-making closer to, and more accountable to, democratically elected bodies (whether local or regional)?
- Are the proposals likely to win public acceptance or support in the areas they affect?
- Do the proposals encourage greater integration, in a way that is consistent with the needs of existing regional or sub-regional partnerships?
- Do the proposals involve active participation in decision taking of people who have much to contribute, but no wish to engage in politics (for example employers, voluntary organisations, academic institutions)?
- Within a region, is there acceptance of a common regional boundary for the relevant functions?

The LGA concludes: 'For local authorities the key aspect of this debate is the need to work towards renewal of their own democratic legitimacy, so that they can assert their strengths as elected bodies, however the regional agenda develops.'

In this regard the new powers to promote the social economic and environmental wellbeing of their areas, and the duty to produce a community strategy involving statutory and non-statutory stakeholders, as well as the local community, is a significant strengthening of their democratic legitimacy.

Although there is a recognition that local and regional government has an important role to play in a complex society of over 50 million people, there is no consensus about the best way forward to strengthen both the sub-national functions, and tackle the 'democratic deficit' that is afflicting the 'missing middle' in English governance.

The debate about what we want from, and what we can hope to get out of, emerging English regional institutions, and their relationship to the long-standing democratic traditions in local government is going to be of considerable political significance in the coming decade. Existing institutions, especially in Whitehall, see themselves as having much to lose. However, if a settlement that enhances local democracy and delivers the crock of gold called 'joined up government', can be arrived at, the quality of life for people in the regions of England could be considerably enhanced.

## References

Cabinet Office (1999), *Modernising Government*, Cabinet Office, London.
Cabinet Office Performance and Innovation Unit (2000), *Reaching Out, the Role of Central Government at Regional and Local Levels*, Cabinet Office, London.
Dungey, J. and Newman, I. (2000), *The Democratic Region*, LGIU, London.
Local Government Association (2000), *Regional Variations*, report of the LGA's hearing on regions, LGA, London.
Mawson, J. (2000), 'Joined up government and the democratic region', in J. Dungey and I. Newman (eds.), *The Democratic Region*, LGIU, London.
New Local Government Network (2000), *Is there a Missing Middle in English Governance?*, NLGN, London.
Regional Policy Commission (1996), *Renewing the Regions: Strategies for Regional Economic Development*, Sheffield Hallam University.
Stott, M. (2000a), 'Regional Manoeuvres', *Town and Country Planning*, vol. 69 (6), pp. 182–83.
Stott, M. (2000b), 'Come clean over local turf wars', *Fabian Review*, vol. 112 (3), p. 23.

# PART III:
# THEMES AND TOPICS

# 7 Regional Transport Strategies: Fond Hope or Serious Planning?

PETER HEADICAR

**Introduction**

The interaction of transport and land use, particularly at regional and sub-regional levels, is arguably one of the most important issues in contemporary planning. It is central to the process of counter-urbanisation which creates serious problems at either end of the urban spectrum. These include at one extreme the decline and under-investment in inner city areas and, at the other, the excess development demands in many dormitory and rural areas – both symptomatic of a spiral of social and economic polarisation. In relation to transport counter-urbanisation fosters the lengthening of trips and the adoption of car-dependent lifestyles, two of the main factors in the massive and continuing growth in car use. It therefore creates settlement and activity patterns increasingly at odds with notions of sustainability, whether defined in environmental or social terms.

These issues have been there to be tackled for the last 40 years (although for the most part collectively they have not been). At the present time, however, two policy developments have altered the context for addressing them in quite a pronounced and distinctive way. They are:

- the retreat from 'predict and provide' as the dominant maxim in national transport policy from about 1994 onwards, and confirmed in the 1998 White Paper (DETR, 1998a) and its implications for transport decision-making at the regional level;
- the Government's proposals for new regional planning arrangements, as embodied in draft PPG 11 (DETR, 1999).

These two developments come together in the proposal for 'regional transport strategies' (RTS) to be prepared and incorporated in future versions of regional planning guidance (RPG).

In this chapter the two developments are commented on individually before going on to discuss the specific issues surrounding the introduction of the RTS concept.

**The Retreat from 'Predict and Provide'**

The change in national transport policy is most clearly evident in the stance taken towards the motorway and trunk road network. The development and improvement of this network had proceeded almost unchanged in its guiding principles since introduced in the late 1950s (responding to, and in anticipation of, increased traffic demands). However, the national roads programme was scaled down significantly from 1993–94 onwards (Department of Transport, 1994) and the incoming Labour Government conducted a fundamental review of the programme itself in 1997–98 (DETR, 1998b). This resulted in more than half of the 156 inherited schemes being deferred or withdrawn. Forty four were referred for the views of Regional Planning Conferences. Many of the most important schemes, including proposals for the M25 and other heavily loaded sections of the motorway network, were made the subject of 'corridor studies' which are currently being initiated by the Government Regional Offices.

These corridor studies have several features which are fundamentally different from the procedures adopted hitherto:

- the adoption of the 'corridor' approach itself, inviting a strategic, long-term view rather than proceeding on a scheme by scheme basis;
- the widening of many studies' remit to include multi-modal options (in which there is no presumption that any elements of the original road proposals will go ahead);
- there is no presumption that forecast travel demands should be accommodated;
- forecast traffic demands may be influenced through the application of user charges, either on the national network itself or through road pricing or workplace parking levies in nearby cities and towns (subject to legislation and, in the case of the local network, to initiation by the local highway authority concerned);
- the incorporation of a range of transport and planning agencies in the arrangements made for steering the studies.

The particular arrangements being developed for the corridor studies are an extension of a principle announced towards the end of the

Conservative administration. This was that trunk road schemes in general would be considered as part of the process of preparing regional planning guidance and not, as before, presented as a fixed input to it (Department of Transport, 1996).

These changes have far-reaching consequences for the treatment of transport within regional planning. In principle they allow for a more comprehensive and integrated treatment of issues arising in connection with transport investment and management in the main movement corridors, although the exact mechanisms for achieving this (discussed further below) remain unclear. They also withdraw implicit support for the car-dependent, counter-urbanisation process referred to earlier.

Land use proposals outside the main urban areas, whether 'plan-led' or 'developer-led', cannot therefore proceed on the assumption that resulting traffic demands will be catered for as part of the long-term trajectory of traffic growth. Nor can they assume that the conditions traditionally associated with the operation of the inter-urban highway network (speed, freedom from congestion and absence of direct charges) will be perpetuated. The changes thus require a fundamental reappraisal of traditional land use as well as transport policies at the inter-urban level – a need signalled in the 1994 revision of PPG13 (Department of the Environment, 1994).

The general retreat from 'predict and provide' is also reflected in two other innovations in the Transport White Paper which are of great importance to regional planning. These are the introduction of Local Transport Plans (LTPs) and the creation of a Strategic Rail Authority.

LTPs were hailed in the White Paper as the 'centrepiece' of the Government's proposals for delivering 'integrated transport' locally. They supersede the former annual Transport Policy and Programme arrangements and have a much wider remit in both content and process, including elements of partnership and public participation (DETR, 2000a).

Local highway authorities have recently completed the first round of these plans which contain their proposed capital programmes for the period 2001/02 – 2005/06. However the effectiveness of any increases in transport supply will be conditioned by the accompanying measures to manage demand, and in terms of demand management authorities cannot hope to function successfully acting alone. Some form of inter-authority agreement at regional or sub-regional level is essential. This applies in three domains:

- to traditional forms of restraint – physical management, traffic regulation, control of parking and parking standards in new

development – in order to achieve coherence on the ground and to overcome fears of displacing trade to neighbouring authorities;
- to the possible new forms of restraint – road user charging or workplace parking levies – and to the use of their revenues (which, to achieve best effect, may need to underpin investment programmes outside the area of the initiating authority);
- to land use policies – to counter the possible migration of developer interest into areas where transport demand management is relatively weak.

LTPs also require an over-arching regional framework if they are to function as credible local planning exercises with significant public involvement. A crucial (and limiting) feature of the LTP system is its continuing link with the local authority bidding process to Central Government for capital expenditure approval. In practice these 'plans' remain bidding documents; they do not provide the interested layperson with a clear and accessible account of what is *actually* likely to happen in their area. This deficiency could be removed in future cycles of LTP preparation if sufficient guidance were provided by the proposed strengthened transport element of RPG. This would need to indicate the priorities for major transport investment within the region, the spatial strategy for travel demand management and the expenditure levels within which individual authorities should prepare their local transport plans.

The second innovation announced in the White Paper – the creation of a Strategic Rail Authority (SRA) – signifies recognition by the Government of the public interest in developing the nation's rail network. This is being reflected in the way the shadow SRA is conducting the first round of passenger franchise renewals, with much greater attention being given to bidders' longer-term investment proposals and not merely to their required levels of financial support. The SRA is, however, answerable to Central Government and its sensitivity to regional objectives and priorities must therefore be in some doubt. At present its timetable for franchise renewals – many of which will cover a period as long as 20 years – is being conducted independently (and in advance of) the multi-modal studies in key corridors and consideration of the enhanced transport strategy element of RPG generally.

The significance of these innovations (LTPs and SRA) is much greater as a result of the unprecedented increases in public investment in local transport and rail announced in the Government's 10-Year Transport Plan (DETR, 2000b). Investment in local transport is planned to increase from a low point of £400m in 1998/99 to £1,900m in 2005/06. Over the same period public investment in the rail network is planned to increase from zero to £2,700m, complemented by a similar amount from the private sector.

## The New Regional Planning Arrangements

The new regional planning arrangements put forward in draft PPG11 (DETR, 1999) are designed to fulfil three main aims:

- to give greater responsibility to regional planning bodies (RPBs), working with the Government Offices and regional stakeholders, to resolve planning issues at the regional level through the production of draft RPG (final responsibility continues to rest with the Secretary of State);
- to improve the planning process followed in the production of draft RPG, with a more streamlined timetable, the introduction of a public examination, a sustainable development appraisal, and more effective monitoring and review;
- to strengthen the role and effectiveness of RPG by adopting a 'spatial strategy' which extends beyond land use issues and which incorporates 'an integrated transport strategy' for the region, and other measures.

The long-term objective of the new form of RPG (DETR, 1999, 3.4) is 'to develop into a comprehensive spatial strategy for the region; ie to set out the range of public policies that will manage the future distribution of activities'.

RPBs are cautioned in how far they move away from an approach which relies upon implementation through 'the planning system'. Interestingly, however, this system is not defined as solely the statutory development planning system with which RPG has traditionally been connected, but includes LTPs also.

RPBs are advised to 'adopt a step-by step approach focussing on the role of RPG in relation to development plans and LTPs and complementing the regional strategies of the RDAs'. Thus although the new RPGs are recognised as relevant to a number of other types of strategy the local transport plans being prepared by local highway authorities are given special treatment. In the proposed monitoring function it is 'development and local transport plans' which Government Offices and RPBs need to ensure are consistent with RPG.

Viewed from a transport perspective the combining of these two forms of planning activity can be seen to reflect one of the dimensions of 'integration' highlighted in the Transport White Paper, namely, with policy sectors outside transport itself. Draft PPG11 views the preparation of an RTS as an integral part of RPG as of 'key importance to achieving this integration'.

In terms of content RPBs are asked to consider a number of items for inclusion in their transport strategies (Table 7.1). Provided the RTS covers most of these it will create the long-term regional framework for structure plans and Part 1 UDPs, local transport plans and for modal operators in developing their plans and programmes. Although RPG covers a 15–20 year period it is seen as important that the immediate five year regional transport priorities are specified to assist the development of these plans and programmes.

**Issues Arising from the Introduction of Regional Transport Strategies**

The aspirations underlying the introduction of RTSs within RPGs are likely to enjoy wide support. Compared with the insularity which has characterised the development of transport programmes in the past the draft of PPG11 and the guidance issued to local authorities for preparing LTPs (DETR, 2000a) represent significant change. Both make valiant attempts to acknowledge the links between different sectors of transport and between transport and land use planning. Both seek to considerably improve the quality of the planning process adopted in relation to transport issues, particularly in the involvement of interests other than those of the public authorities directly responsible.

However, support for these aspirations does not disqualify debate on the mechanisms planned for achieving them. On these depend the Government's own hopes for devolving responsibility to RPBs, for wider public involvement and for the actual delivery of more integrated policies and programmes.

**Table 7.1: Items for RPBs to Consider Including in their Transport Strategies**

- Regional priorities for transport investment and management (all modes), including local authority roads of regional or sub-regional significance

- A strategic steer on the role and future development of railways, airports and ports in the region, consistent with national policy

- Guidance on measures to increase transport choice, including the better integration of rail and bus services

- Public transport accessibility criteria for regionally or sub-regionally significant levels or types of development as a guide to the location of new development and the provision of new transport services or infrastructure

- Advice on the approach to be taken to standards for off-street parking provision

- Guidance on the strategic context for demand management measures such as congestion charging and workplace parking levies

*Source:* Draft PPG 11

Three issues will be considered here:

- The form of governance within which the RPG/RTS process is to be conducted;
- The relationship between development plans and LTPs as the first example of RPGs embracing a wider 'planning system';
- The relationship between the different sectors of transport to be represented in Regional Transport Strategies.

*The Form of Governance*

This is a subject which is relevant to all aspects of regional planning. However, transport, or more specifically, land use/transport integration, does seem to pose particular problems which deserve to be highlighted.

By its nature transport is directly concerned with 'physical connectivity'. Hence the nature of the *geographical units* within which planning is conducted above the level of individual local authorities is of great importance. As far as possible the units need to reflect the divisions into which the major day-to-day flows of movement naturally fall. In this context the present regional units are in some cases either unnecessarily large (for example the South West) or unrealistically small (London and the South East). The former, though unwieldy, is not critical; the latter is. (This is putting aside matters of precise boundaries and the treatment of cross-boundary issues which will arise whatever the scale of unit.)

A particular difficulty arises with the 'Greater South East'. The more arrangements here might be modified to accommodate issues of a genuinely regional scale, the further they become removed from being able to give physical expression to its consequences at the sub-regional level (in effect the corridors radiating from Greater London into the surrounding administrative regions). Draft PPG11 acknowledges (DETR, 1999, 3.6) that '...In some regions or parts of regions it will be appropriate for RPG to contain sub-regional strategies because of the specificity of the issues and relationships at that level that cannot be resolved if left to individual authorities, but which are not general enough to be addressed at the regional level'.

However, this does not seem an adequate response to conditions in the Greater South East where *every* part (i.e. major corridor) seems in need of such a sub-regional strategy, especially where the introduction of unitary councils has fragmented the former county units.

There is some irony in the fact that, as far as spatial planning in the Greater South East is concerned, the most important scale in terms of achieving integrated land use/transport planning (i.e. the sub-regional corridor) is the one which is not formally represented in administrative arrangements! Obviously one needs to take care not to view national proposals for regional planning from too London-orientated a perspective. Nevertheless the problems faced in trying to devise adequate arrangements for the Greater South East surely constitute more than a little local difficulty and challenge the notion of a regional planning structure which can be applied uniformly across England.

There is then the question of the *form of government* to be developed for the regional units, however they are drawn geographically. The Labour Party's initial (and always controversial) preference for directly elected regional assemblies seems to have slipped down the agenda since arriving in government. Experience to date with the *ad hoc* arrangements for inter-authority collaboration at regional level does, however, pose doubts about their efficacy. In fact can it be considered reasonable to expect a collection of members representing autonomous local authorities *ever* to approve a strategy which imposes serious policy strictures upon these authorities' individual actions and which incorporates hard decisions about winners and losers as far as major investment is concerned? This is not an issue unique to transport. However, our assessment of local authorities' Provisional LTPs (Oxford Brookes University, 2000) has highlighted the vacuum surrounding delivery of the Government's more 'radical' policies in the absence of such a strategic framework.

*The Relationship between Development Plans and LTPs*

Reference in Government guidance on Regional Planning to development plans and LTPs as a single 'planning system' may be a worthy aspiration but it does not accord with current realities. Local authorities' actions as highway authorities are not subject to the same disciplines in terms of conformity with national and regional policy as are their actions as planning authorities. Their policy documents (now LTPs) are not subject to public examination chaired by an independent inspector. There is no equivalent of the 'call-in' procedure whereby individual local decisions which raise or challenge issues of national policy can ultimately be referred for ministerial decision. (In effect the approval or otherwise of highway authorities' transport expenditure programmes in LTPs represents a 'back-door' process whereby national policy is exerted.) The mechanisms simply do not exist which would enable an interested outsider to have the same degree of confidence that policies adopted at regional level will be carried through locally.

These differences which exist between development and transport planning systems generally are compounded by the splintering of functions at local authority level. No regional planning strategy can or should provide the degree of detail which resolves more local issues of land use/transport integration. How then are conflicts of interest between individual localities, authorities and policy sectors to be resolved? The Structure Plan process would seem the obvious forum. However, the arrangements so far developed for LTPs seem almost deliberately to negate this. Remarkably neither the *timing* of LTP submissions, nor in many cases the *geographical units* for which they are being prepared, have so far been geared to coincide with Structure Plans.

Thus we have the anomalous situation of non-statutory, shorter-term, topic-specific LTPs, characterised by relatively weak opportunities for public involvement and scrutiny, being progressed in advance of the statutory, longer-term, openly scrutinised, comprehensive spatial strategies for an area – and in many cases by a different set of authorities! The Government's guidance for LTPs urges local highway authorities – their authors – to ensure 'consistency' with development plans. It also invites local planning authorities to amend their development plans to match LTP provisions where there is a discrepancy. But what encouragement does this offer for genuinely integrated planning? Where in all this do other stakeholders see the opportunity for holistic public discussion and scrutiny?

The desirability of bringing together development and transport planning arrangements below regional level presents formidable difficulties – principally sacrificing the autonomy and relative speed of delivery available to local highway authorities to the complications arising from the broader remit and greater public scrutiny allowed for in development plans. If LTPs were to be linked directly to the Structure Plan process for example 'delay' would be created between adopting regional transport strategies and translating their investment proposals into schemes 'on the ground'. Current arrangements maintaining the independence of LTPs from the statutory development planning process are therefore explained away officially with the assertion that once the new form of RPG is in place the content of LTPs can be derived directly and unproblematically. Would that it were so simple!

The new form of RPG cannot honestly be presented as the apex of integrated land use/transport planning in a region whilst the mechanisms beneath it remain so disjointed. It is not that integrated planning (through to implementation) is impossible where there is sufficient political will and consensus between the constituent authorities and other stakeholders. But the true test of a 'planning system' must surely be its performance in the more demanding situations where there is not such will and consensus.

*The Relationship Between the Different Transport Sectors in RTS*

The discussion above has highlighted the problems likely to arise because of the different territorial and functional responsibilities of local authorities engaged in various elements of land use/transport planning. However, aside from these differences, local authorities share important features within the regional planning process. They are public bodies, accountable to local electorates, who together have the lead role in drafting RPG/RTS. They have no direct interest in transport as an industry. More than anyone else they can be expected to view transport in terms of its contribution to wider economic, social and environmental objectives.

The RTS, however, requires the participation of many other bodies who do not share these characteristics. Draft PPG11 (DETR, 1999, 6.5) acknowledges that '...the successful implementation of the transport strategy will depend upon the co-operation of a large number of different organisations in both the private and public sectors'.

Within the public sector the Highways Agency and the SRA are key players. Unlike local authorities they are single purpose agencies without any form of local accountability. They are agents of Central Government. This is significant not only where their actions within a region relate to some element of the *national* transport system, but also where they may use their control of essentially regional assets and expenditures to advance the preferences of national government. In the current era of political hegemony the potential policy differences between central and local government do not receive much attention, but reference back to the era of the Greater London Council (GLC) and Metropolitan Counties reminds us how critical this issue may become. Again this begs the wider question of regional government and its potential for receiving devolved functions currently discharged nationally. In this regard it will be particularly interesting to compare experience with regional transport strategies either side of the Scottish border.

Similar comments apply to the role of the RDAs. Their combination of independent status and sectoral focus means that, while they will have an interest in influencing the RTS, they cannot be expected to abide by elements which they would regard as constraints. Treatment of the link of RDA strategies with RTS specifically in draft PPG11 (DETR, 1999, 6.6) merits quotation for its exceptional blandness: 'As with the rest of RPG, the RDA will be a key stakeholder and it will be important that the regional transport strategy and the RDA's strategy pursue complementary approaches. The RPB will need to take account of the transport implications of the RDA's proposals and similarly the regional transport strategy should assist the RDA in the development and implementation of its strategy.'

Like the link between local transport plans and development plans discussed earlier these sentiments do not preclude a synergetic combination if circumstances are favourable, but they do not offer much assurance if they are not.

In theory at least, if the issues are seen as important enough, reconciling differences between the various public bodies at English regional level can be the subject of ministerial intervention. Public bodies, whatever their individual remit, can also be charged with the need to have regard to the wider public interest, including the merits of open public debate. Private bodies, by definition, are not subject to these constraints. They can only be expected to contribute to the formulation and implementation of an RTS insofar as they see it as furthering their particular interests or insofar as the regulatory framework within which they operate requires it. This is critical now that public transport operation, rail infrastructure, ports and airports have been transferred almost entirely to the private sector.

Draft PPG11 states that 'these bodies should be invited to outline their current plans and proposals for the region and to discuss how these support the sustainable development objectives for the region'. This suggestion begs two questions which seem so fundamental (and so obvious) that it seems almost gratuitous to raise them here. What happens if they do not wish to publicise their intentions and – even if they do – how does one proceed if these do not support sustainability objectives?

A central tenet in the privatisation of transport industries was the introduction of competition within and between modes. Information about future intent is a key resource and is unlikely to be released where it would give advantage to a competitor or where it might give advance opportunity to groups who might be expected to mount some form of public opposition campaign. We do not, for example, expect rival supermarket chains to come forward and discuss their strategies publicly before they submit individual planning applications. However, even this analogy understates the potential difficulties to be faced in dealing with the transport industries because, in the case of transport developments, there is not normally the ultimate sanction of refusing planning permission to encourage them to engage in public debate and to try and win public support. Confidentiality aside, why should private firms put themselves in a position where they commit themselves to lengthy, tortuous and potentially hostile public debate (where the agenda is likely to be set by others) if they do not have to?

Reference to sustainable development objectives in RTS is particularly pertinent given the significance of transport to environmental damage and social polarisation. However, private transport industries are under no obligation to sign up to 'sustainable development' and certainly not to local authorities' particular interpretation of it in their areas. There can be few more glaring contradictions than between these industries' core interest in 'growing the market' and the injunction to planning authorities to 'reduce the need to travel'. Even the fact that authorities may wish to promote the use of public transport in certain situations does not mean that these will be where operators see the greatest commercial opportunity. Yes, there will be situations where interests coincide and partnerships can be developed. However, that is not the same as an integrated regional strategy. As with development planning the acid test is the handling of situations where the public interest and the interests represented in the market do *not* coincide.

These comments should not be read as a lament for the passing of public ownership. Rather, if there are public interests to be safeguarded – whether in the operation of transport industries or in debates over their future development – then the transfer of ownership must be accompanied by adequate regulation. As yet the Government has shied away from the political discomforts of such logic as far as spatial strategies are concerned.

## Conclusion

Those who support the Government's objectives, and who wish to see the role of public planning enhanced, must be increasingly concerned at its unwillingness to address the issues described in this chapter. Intractable as they undoubtedly are, to will the ends without the means is but a recipe for ineffectiveness and disillusion. Fond hopes need to mature into serious planning.

## References

Department of the Environment (1994), *Planning Policy Guidance Note 13: Transport*, HMSO, London.
Department of the Environment, Transport and the Regions (1998a), *A New Deal for Transport: Better for Everyone*, Cm 3950, The Stationery Office, London.
Department of the Environment, Transport and the Regions (1998b), *A New Deal for Trunk Roads in England*, DETR, London.
Department of the Environment, Transport and the Regions (1999), *Draft Planning Policy Guidance Note 11: Regional Planning*, HMSO, London.
Department of the Environment, Transport and the Regions (2000a), *Guidance on Full Local Transport Plans*, DETR, London.
Department of the Environment, Transport and the Regions (2000b), *Transport 2010: The Ten-Year Plan*, DETR, London.
Department of Transport (1994), *Trunk Roads in England: 1994 Review*, HMSO, London.
Department of Transport (1996), *Transport: The Way Forward*, Cm 3234, HMSO, London.
Oxford Brookes University (2000), *Towards Better Local Transport Planning: The Performance of Provisional Local Transport Plans*, School of Planning, Oxford.

# 8 Re-Shaping Regions, Re-Ordering Utilities: Competing Development Trajectories

SIMON MARVIN

**Introduction**

This chapter examines the relations between regional strategies and the management of utility and infrastructure networks (on these networks, Guy *et al.*, 2001). It explores the linkages between the multiple constructions of a regional future and the style of utility network that would underpin these views of the future. The interface between regional futures and sociotechnical futures is explored through the window of three regional strategies in the North West of England. This analysis is designed to highlight a key controversy that is poorly tackled in the wider debate about regional development and infrastructure – how competing visions of regional futures shape networks.

Over the last decade the gap between central government and local authorities has been filled by a complex array of regional organizations – the most significant are regional development agencies, government offices and regional assemblies, but there is also a wide range of non-statutory agencies developing their own notions of regional development. Within the North West there is a multiplicity of agencies whose interconnections are sometimes captured in complex wiring diagrams. There has been a tendency to bemoan the complexity of regional governance, to express concern about the lack of organisational and political clarity, and to argue the need to develop more 'integrated' and 'joined up thinking' between different regional viewpoints.

This chapter, however, takes a slightly different position. Before we can begin to argue for better integration there is a need to develop a conceptual framework which is sensitive to the development trajectories implied by multiple regional strategies and which is able to identify the resonances and dissonances between different viewpoints. Without an understanding of the affinities between competing views of the future it is difficult to know what would be 'joined-up'.

The chapter is organised in four sections. Firstly, the chapter develops a conceptual overview of the changing relations and questions raised by the development of regions, and infrastructure. Secondly, the most substantial section of the chapter looks at the re-configurations to infrastructure networks, such as energy and water utilities, embedded in three regional strategies. Each strategy is produced by different social organisations, focused on particular regional problems and generates quite different visions of infrastructural change. The section highlights how each viewpoint shapes the development of complex socio-technical infrastructure networks along different development trajectories. Thirdly, the chapter then compares and contrasts the viewpoints illustrating the development of multiple levels of infrastructure re-ordering, and points to the absence of a context for bringing the different approaches together within a region. The chapter concludes by calling for the development of a new conceptual framework that is sensitive to the multiple re-orderings of infrastructure. The challenge is to develop a 'language' that can capture these and identify the contradictions and opportunities between different futures. If we can do this then we might begin to develop an understanding of what we can start to 'join up' or, alternatively, how multiple visions of the future may be able to co-exist.

## Conceptual Framework: Shaping Regions/Ordering Infrastructure

This chapter constructs an understanding of the relations between the emerging regional development agenda and the wider debates about the infrastructure and utility needs of the region. New tiers of regional institutions are creating quite different visions of a future North West. Within the last two years a range of regulatory, voluntary and environmental groups, government agencies and groupings of local authorities have produced their own regional strategies and plans setting quite distinct pathways towards a sustainable regional future. Yet there has been little attempt to assess the significance of these regional strategies for the future development of utility and infrastructure networks or the affinity between the different strategies. This chapter develops an understanding of the implications of the different pathways towards a sustainable regional future and the implications for the management of utility and infrastructure networks.

There is a wide range of different strategies and plans that we could examine. Here we focus on plans and strategies that exemplify quite different assumptions and visions – the Regional Economic Strategy, the North West Climate Change Report and the Regional Planning Guidance. Each strategy reflects different social interests, styles of problem definition, planning methods and policy recommendations. Yet the key is that they all claim to represent a vision of a 'sustainable' North West. Taking each strategy, the chapter focuses on a number of themes that cut across all three strategies. What is the regional problem the plan is addressing? What vision of the North West futures is implied by the strategy? And finally what style of infrastructure network is implied by the strategy? The chapter builds a typology (Table 8.1) of different views of the region's future, the social and technological assumptions built into particular visions, and the development of a conceptual framework for assessing the implications for the management and development of infrastructure and utility services.

## Regional Strategy – Networks as 'Infrastructural Assets'

The first window we look through on to re-ordered infrastructure is 'England's North West: A strategy towards 2020' produced by the North West Development Agency (NWDA, 1999). The strategy was originally conceived as much more than a regional economic strategy – it attempts to build a social and environmental framework for the future development of the North West. The strategy has a 20-year timescale and was prepared by a team at the NWDA, including secondees from local authorities in the region with substantial support from consultants. The strategy is the result of major and widespread consultation across the region, informed by commissioned research. Work started preparing the strategy in February 1999, the draft was published in July – in total there were 500 written submissions, 16 panel hearings, 12 consultation meetings and 31 private sector working groups.

The strategy seeks to balance social, economic and environmental objectives. The prime objective is to develop the economic competitiveness of the region, but this should support sustainability and social inclusion. 'Our vision, which we commend to the North West and to Government, is a region which: Attracts and retains the skilled and talented, Brings everyone into the mainstream of community life, Nurtures its environment, heritage and culture, Kindles creativity, innovation, and competitiveness, Transforms its image, Strengthens its infrastructure, And is naturally in the shortlist for new investments' (NWDA, 1999).

**Table 8.1: Three Views of Regional Infrastructure**

|  | **Regional Strategy** | **Regional Climate Change** | **Regional Planning Guidance** |
|---|---|---|---|
| Organisation | Development Agency | Climate Change Group | Regional Assembly |
| Purpose | Increase economic growth | Prepare for climate change | Shape spatial framework |
| View of infrastructure | Support the competitiveness of the region | Create a more resilient regional infrastructure | Build a more self-reliant regional infrastructure network |
| Key features | Infrastructure as:<br><br>• Growth sector<br>• Supporting economic efficiency<br>• Affecting image | Infrastructure as:<br><br>• Vulnerable<br>• Strengthened<br>• New technologies | Infrastructure as:<br><br>• Autonomous resources<br>• Balanced<br>• Role in demand management |

The strategy designed to achieve this vision is organised in four related themes, which are: Business and Ideas; People and Communities; Infrastructure; and, Image and Environment. In each theme, three objectives are identified, together with action points to deliver those objectives. Priorities for early action are drawn from the action points. These comprise the Action Programme for the Agency and partners in the first 12 months.

Identifying a distinct role for utilities and infrastructure networks in the strategy is problematic. The strategy provides an overview of key strategic priorities and will be supplemented by other documents that then build upon these priorities in much more detail. There is also the problem that the section dealing with infrastructure mainly focuses on communications − transportation in particular − and has relatively little to say about energy, water and telecommunications. However, by carefully reviewing the different sections of the strategy it is possible to begin to build a notion of the overriding role of re-configured infrastructure as a 'competitive asset' for the region. The dominant story line is the different roles that infrastructure can and should play in development of a competitive region.

*Growth Sectors and Infrastructure Services*

A number of new or enhanced roles for infrastructure services are implied in the section on Business and Ideas. Firstly, the strategy identifies seven target growth clusters including environmental technologies. Many of these technologies are closely associated with energy production and consumption, water treatment and waste disposal. Consequently the utilities sector may have a key role in the demonstration, development and applications of such technologies in their networks. Alternatively the strategy could create a context for intermediaries (consultancies, energy service companies, charities and so on) who may have a commercial or social interest in inserting such technologies into users' premises via infrastructure and utility networks.

Secondly, there is another set of established industries that are identified to complement the development of the new growth clusters. A target here is the energy sector (including nuclear energy and reprocessing), which provides 15,000 jobs in the region. Although mainly focused around the nuclear sector there will presumably be strategies to develop the energy sector within the region creating opportunities for utility sector involvement.

Finally the NWDA will develop policies to build business excellence within the region. Among the six strands of this activity is: Environmental best practice – work with local partners to enable the region's businesses, especially smaller firms, to gain business benefits from environmental best practice, comply with requirements of environmental legislation, and become more efficient in their use of resources and energy (NWDA, 1999, pp. 23–4).

This suggests measures, presumably led by intermediaries and/or utilities, to improve the efficiency and quality of water and energy consumption and reduce levels of waste generation in businesses. Clearly then the implication is that utilities should support business development – partly through the development of the new sector of environmental technologies, expansion in existing energy industries and through improvements in efficiency of resource consumption in existing businesses.

*Infrastructure Supporting Development*

The second theme developed in the strategy is embedded in the section on Infrastructure. While this has relatively little to say about energy, water and even telecommunications there are a number of assumptions about the role and quality of energy and water services in relation to servicing newly built development. These take three forms. Firstly, there is the encouragement of the re-use of infrastructure within existing built up areas and brownfield sites. Presumably this means a focus on exploiting underused and spare capacity on existing networks with no or insufficient demand. The implications of this approach would be to steer development to cold spots above underutilized areas of the networks. The second proposal is for sustainable growth through new development in the Southern Crescent. (The Southern Crescent is a sub-region covering part of Merseyside, north Cheshire and Southern Manchester identified as a potential growth area in the Regional Strategy.) Although the concept of sustainable growth is not developed in any detail, the implications for infrastructure might be such developments as highly energy and water efficient buildings, combined heat and power or renewable energy production in new development sites or the application of novel environmental technologies. Finally the strategy identifies the need for infrastructure support for a portfolio of strategic sites in the region. There is very little detailed discussion about how this will be achieved or what issues will have to be addressed.

*Infrastructure and Image*

A number of these issues are hinted at in the final section on Image and Environment. In this theme environmental quality and the region's external image are closely linked with the objective of developing the region as a site for inward investment and tourism. Investment in environmental assets is viewed as having wider economic benefits in promoting efficiency, image and quality for the region. Infrastructure networks are caught up in this vision in a number of different ways, as owners and managers of significant landscapes, as providers of services and as marketers of key places. The strategy focuses on a number of these different dimensions. 'The region is heavily dependent on non-renewable energy. Businesses need to improve their energy efficiency and waste minimisation practices' (NWDA, 1999, p. 52).

The most concrete objective related to utilities is to 'promote high standards of architecture, design and energy conservation and encourage the development of renewable energy sources'. The NWDA signal their intention to promote development that reaches high standards of energy conservation and management and assess the scope for developing alternative energy sources in the region. Thus the development of highly resource efficient and more autonomous buildings is linked to the creation of the region's environmental image and quality, although the precise nature of these linkages is not specified.

Overall the strategy identifies a limited number of objective and performance criteria, but the only ones that relate to infrastructure are targets to reduce the proportion of waters classed as poor and an increase from 6 to 13 in 2006 in the number of new office development meeting Building Research Establishment best practice standards. Clearly there is relatively limited focus on infrastructure within the current framing of the strategy and more detail is likely to emerge when the related documents and sector strategies are developed. However, there are currently enough references and concepts to build a picture of how infrastructure might be re-ordered.

The key role is the development of technologies that can build new economic sectors or growth in the North West and the diffusion of these environmental technologies in small and medium enterprises to improve efficiency of resource and energy use. These imply that utilities could themselves become active agents in the development of such technologies and the expansion in autonomous buildings and renewable energy production. Alternatively a series of intermediaries could focus on the development and diffusion of technology. The key concept though is the role of infrastructure in promoting 'competitive advantage' for the North West.

## Climate Change – 'Resilient Infrastructure'

The next window focuses on the study 'Everybody has an Impact: Climate Change Impacts in the North West of England' (North West Climate Group, 1998a) initiated by the North West Regional Assembly (NWRA) to examine the impact of climate change. The report identifies the beneficial and detrimental effects on a series of landscape domains and economic sectors. Implicit within the analysis are different views of how infrastructure networks would be affected by climate change and the types of management changes that would be needed to deal with and reduce these potential impacts. Perhaps the defining notion of infrastructure that emerges from this vision is the notion of infrastructure that needs to be increasingly resilient to respond to the threats produced by climate change.

The context for the report is the huge increase in the burning of fossil fuel worldwide over the last 50 years, leading to increased global carbon emissions. Carbon dioxide levels in the atmosphere have increased by 30% over pre-industrial levels. The report argues that there is a need to look at the impacts of climate change at the regional level and assess how institutions will have to come to terms with the expected changes. The report argues that climate change presents an unprecedented challenge for the region. 'Climate change is already a fact of life in the North West of England. Sea levels at Liverpool have been rising by 1 cm per decade and average temperatures taken at points such as Manchester Airport have been increasing consistently since the 1960s. Levels of winter rainfall have been increasing, and recent flooding and periods of drought only serve to underline the very real need to prepare for an uncertain future as climate change continues' (North West Climate Group, 1998b, p. 2).

The study is the first regional integrated assessment of the impact of climate change for the United Kingdom. It focuses on how climate in the past and future shapes socio-economic practices, industry and business affairs, land management and social life in the North West region. The report highlights significant areas for action and points the way forward for future areas of research. The starting point is that we know that certain changes are taking place in the climate and that there is a need to help people, businesses, and agencies to plan for the changes and be prepared to deal with the impacts. So the study presents a 'vision' of a future North West region to institutions to help them develop a response at two levels, by reducing the level of $CO_2$ emissions and by starting preparations for climate change.

UMIST and four other universities carried out the study for the North West Climate Group. The North West Climate Group comprises: North West Regional Assembly; Sustainability North West; North West Water; Environment Agency; National Trust; Government Office for the North West; AXA Insurance; and the UK Climate Impacts Programme. The methodology combined expert judgment; stakeholder assessment; qualitative and quantitative scenario construction and assessment methods; literature review. The research team spoke to 148 individuals from 65 companies and 52 public sector agencies, policy and research organisations based in the region. The study develops scenarios for the 2020s, 2050s, 2080s and projects four alternative levels of climate change with low, medium-low, medium-high and high sensitivity to the effects of greenhouse gas emissions. The scenarios (NWCG, 1998b, p. 3) suggest that:

- The rate of future warming in the North West varies from 0.1°C to 0.3°C per decade (1 – 3°C per century).
- Annual rainfall will increase for all UKCIP scenarios, by between 3% and 5%. Winter rainfall over North West England will increase by between 6% and 14% by the 2050s, whilst a decrease is forecast for summer rainfall varying from 1% to 10%.
- The record-breaking summer of 1995 will, under the UKCIP medium-high scenario, become the norm in the 2050s, while an extreme summer in the 2050s will be 4.7°C higher than the 1961–1990 average.
- Some winters in the 2050s may be up to 3.5°C warmer than the 1961–1990 average.
- The UKCIP scenarios give a potential sea level rise of between 12cm and 67cm by the 2050s, with the medium-high scenario resulting in rates of sea level rise more than double those observed at Liverpool over the last 100 years.
- Days with minimum temperatures below freezing are likely to be reduced by 65% by the 2050s under the medium-high scenario.
- Days on which the temperature exceeds 25°C will double.

The report looked at the impact of these shifts on five 'landscape domains' (coastal zone, rural lowlands, rural uplands, urban fringe and urban core) and on the key economic sectors in the region (chemicals, manufacturing, buildings, energy, tourism and leisure, insurance, water, transport, cultural, public policy). Here we focus on the implications for the water and energy infrastructures.

*Vulnerable Infrastructure*

The first set of issues focuses on the increased vulnerability of infrastructure networks in the context of climate change. These are particularly strongly felt in the water sector. 'Changing patterns of rainfall are of particular concern to the water industry in the North West with periods of prolonged drought, resulting in reduced availability of water for customers' (North West Climate Group, 1998b, p. 6).

Overall the context within which the water network will be operating becomes much more stressed, increasing the vulnerability of the networks (North West Climate Group 1998a, p. 23). Over the last few decades the industry has had to cope with large variations in rainfall patterns. This provides useful experience for dealing with variations that might be expected with climate change. The problem is that predictions of what will happen can only be made with low confidence and relate to average conditions over a whole year. Because water supply availability is most sensitive to the intensity and duration of drought these predictions cannot be used as a basis for long-term planning. There is, therefore, considerable uncertainty about how the system will cope with climate change. There are also additional problems that are likely to be accelerated by climate change – these include water discolouration that increases treatment costs, the flushing out of pollutants collected in droughts affecting water quality, and more intense rainfall that can flood sewage systems leading to foul flooding of water courses.

Climate change seems to have a relatively minor impact on the supply and distribution of energy in the North West (North West Climate Group, 1998a, p. 33). Increasing temperatures are likely to reduce demands perhaps by 1.5% in 2050, but this does not take into account rising demand due to increased use of air conditioning. Increased frequency and severity of extreme weather events could increase the threat to the electrical distribution network. High winds during the storms of winter 1997 disrupted NORWEB's and MANWEB's distribution network, with power not restored to remote areas for some days. The drying out of soils may also reduce the cooling of underground electricity cables, increasing their vulnerability to failure.

*The Need for a More Resilient Infrastructure*

Increased stress means that the vulnerability of the networks could increase unless efforts are made to build the resilience of the networks. There are a number of ways in which this could be done. One response is to build greater 'headroom' into the planning process to cope with the assumed threats. 'Should climate change result in increased severity and frequency of droughts there will be significant challenges for water supply. Further water conservation, leakage reduction and efficiency measures, as well as some new water source development, will be necessary to ensure the continuing reliability of water supplies' (North West Climate Group, 1998a, p. 23).

Waste treatment facilities have been designed to cope with three times the projected demand placed on the system. This safety margin may need to be revised if changed rainfall patterns were to increase the load on such plant during the winter months. The capital costs of increasing safety margins could be very significant, although they could more easily be built into new construction. Electricity companies may have to build in climate headroom when making major capital investment in distribution networks, to build greater resistance to high winds. Companies may also have a key role in reducing the level of $CO_2$ emissions through targeted energy efficiency and conservation programmes. There is also the possibility that specialized industries could benefit through the development of renewable energy products like photo-voltaics, which reduce $CO_2$ emissions and which, through careful siting within the distribution network, can increase the resilience of the whole system.

Overall the implications of climate change are that infrastructure networks may need to be reconfigured to cope with increased threats and that vulnerability could be reduced through efforts to build greater resilience into the networks. This might include a number of complex social and technological responses, such as reshaping planning processes to build greater 'headroom' for unexpected threats, increasing the technological resilience of the network through the incorporation of more autonomous and decentralised technologies, or alternatively preparing users for the possibility of service interruptions and enhancing their role as energy and water savers in times of stress.

**Regional Planning Guidance – 'Self-Reliant Infrastructure'**

The regional planning guidance develops a spatial strategy which provides the broad strategic spatial framework for the preparation of development, transportation, and waste management plans, and informs the programmes of other agencies including the development agency and private infrastructure providers. The spatial development framework has two core aims – the creation of sustainable patterns of growth and change across the region and the concentration of growth in regional centres and towns.

There are two principles that will be applied to new development, each with important implications for the development of infrastructure networks. The first principle is 'economy in the use of land'. This means that new development should be assessed against a sequential approach that ensures 'effective use of existing buildings and infrastructure', 'use of previously developed land' and then 'development of previously developed land' which is well located in relation to infrastructure (NWRA, 2000, p. 5). Taken together these policies would suggest that effective use must be made of existing infrastructure networks and that development should not be located in areas that would make significant new demands upon infrastructure services. Such a development strategy would need to be steered by the identification of those sites that have cold or underutilised infrastructure – brown field and regeneration zones.

The second principle is 'quality in new development', with policies that: '...Encourage innovative design to create high quality living and working environments, especially in housing terms, with more efficient use of energy and materials, and more eco-friendly and adaptable buildings including sustainable drainage systems' (NWRA, 2000, p. 6).

Here the overriding principle is that new development should be designed in such a way that it makes lower demands on infrastructure networks than conventional buildings do. This would require that more decentralized energy and water efficiency / conservation technologies are incorporated into new buildings to reduce their demands for infrastructure services.

The next set of policies that shape the development of infrastructure networks are written into the section headed 'Prudent management of environmental resources'. There are a number of proposals all oriented around the concept of building a more self-reliant infrastructure network. In the case of water, new development should only take place where adequate water resources already exist or 'could be affordably developed without environmental harm, with water saving measures and water efficient equipment incorporated wherever possible' (NWRA, 2000, p. 47). The assumption is that water networks should be re-used and, in cases where this is not possible, buildings should be designed to reduce the demands they place upon infrastructure networks. The concept of self-reliance is much stronger in the section on renewable energy and energy efficiency. Policy ER11 states that local authorities should: 'Support local initiatives and proposals for renewable energy installations that promote self-sufficiency in energy generation and use. Development plans should:

- ensure that development minimizes energy use through careful and imaginative location, design and construction techniques;
- encourage the use of energy efficient technologies such as Combined Heat and Power plants and eventually photovoltaics in major new development and redevelopment; and
- identify areas of search with criteria based policies for renewable energy development' (NWRA, 2000, p. 50).

*Self-sufficiency*

Overall we can see that the thrust of policies shaping the development of infrastructure is based around the concept of greater 'self-sufficiency'. This expresses itself in three ways. Firstly, there is the re-use of existing infrastructure – with a premise that new investments are made as a last resort. Secondly, efforts are to be made to increase self-sufficiency in energy and water supplies through the development of embedded energy production. Finally, new developments should make fewer demands on networks through the incorporation of water and energy conservation/efficiency technologies within new development.

Each of these shifts would dramatically reconfigure the social and technical organisation of infrastructure networks. Initially utilities would have a significantly enlarged role in identifying the sites that had spare capacity (cold spots) to ensure that development strategies focus growth pressure in these areas. But more significantly there would be major changes to the development process – utilities, buildings designers, and constructors would have to insert a much more complex set of technologies into the development process and change social practice in building management to reduce the demands that new development placed on infrastructure networks.

## Multiple Levels of Infrastructural Re-ordering

By focusing on infrastructure networks we have begun to build an understanding of how three of the most significant regional strategies would re-order the social and technical organisation of these networks. Each strategy has embedded within it important assumptions about how infrastructure networks would have to be re-ordered to support the new context developed in each strategy's vision of the future. Table 8.1 provides a synthesis of the view of infrastructure taken inside each strategy and a summary of how the infrastructure network would be re-ordered. While the summary clearly identifies a particular trajectory of infrastructure development, the interesting issue is to identify the affinities between these different pathways.

There is one area where powerful resonances between the strategies can be identified. Across all the strategies a common theme is the insertion of new environmental technologies into infrastructure networks, particularly at the stage of newly built development or refurbishment. Although each strategy had a different rationale for the insertion of these technologies – an economic dimension oriented around a new growth sector and business efficiency, an environmental rationale to reduce demand on networks and a resilience rationale to reduce the vulnerability of networks to climate change – these viewpoints appear to be mutually self-supporting.

However, what is almost entirely missing from these strategies is any understanding of the social context within which these technologies will be inserted into buildings or any identification of key drivers (commercial, social, regulatory and so on) that might accelerate their take-up. So for instance – as we have already hinted above – incorporation of these technologies will require quite significant shifts in the behaviour of planners, architects, utilities, building services and so on. There is relatively little understanding of the role that the commercial, public and private sectors can play here or how the changes will be managed and what incentives will be created to accelerate different styles of infrastructure provision.

Turning to the dissonances between the strategies, the most significant discontinuity is between the emphasis on re-use of brownfield sites within the Regional Planning Guidance and the opening up of a new greenfield development trajectory in the Southern Crescent in the regional strategy. In one sense both the regional strategy and planning guidance support the re-use of infrastructure and development on brownfield sites and both are mutually self-supporting in demands for insertion of new technologies in new development. Yet it is less clear how the greenfield issue will be resolved. The purpose here is not to say which is right or wrong, but what is significant is that we seem to lack an arena in which the broader relations between different types of regional strategy can be brought together, examined and the level of affinity assessed.

The challenge then is not simply to develop 'joined up thinking' or closer 'integration', but to develop a much more transparent and negotiative process that is sensitive to the synergies and discontinuities between regional strategies. So for instance looking at infrastructure and built development, we need an approach that brings the different elements of the region together, examines how new environmental technologies can be inserted into the development process and explores the implementation pathways for doing this.

In the case of the broader spatial development strategy, a more complex shift is needed, which pays careful attention to the context and the direction of change implied by different regional strategies. While we have focused on one issue the challenge is to extend this analysis and look at other thematic issues to build a more careful understanding of how competing strategies define the problem, propose particular solutions and generate different social visions of the future. In those areas where there is agreement on the problem definition and the solution (as in the case of environmental technologies), a different process is needed to accelerate change. But where strategies do not agree on the solution (in the case of strategic spatial development) we need an alternative approach for conflict resolution. However, until we develop a process that can map these affinities, it will be difficult to generate a debate about how the tensions are to be resolved.

## References

Guy, S., Marvin, S. and Moss, T. (2001), *Urban Infrastructure in Transition*, Earthscan, London.

North West Climate Group (1998a), *Everybody has an Impact, Changing by Degrees, The Impact of Climate Change in the North West of England*, Technical overview, Sustainability North West, Manchester.

North West Climate Group (1998b), *Everybody has an Impact, Climate Change Impacts in the North West of England*, December, Summary Report, Sustainability North West, Manchester.

North West Development Agency (1999), *England's North West, A Strategy Towards 2020*, NWDA, Wigan.

North West Regional Assembly (2000), *People, Places and Prosperity, Draft Regional Planning Guidance for the North West*, July 2000, NWRA, Warrington.

# 9 Promoting Sustainable Development through Regional Plans: The Role of the Environment Agency

HUGH R HOWES [1]

## Introduction

The Environment Agency for England and Wales is a powerful regulatory agency, yet one which depends on other regimes such as regional and local development planning to achieve some of its aims. This chapter examines the recent changes in the approach taken by the Agency in influencing the policies for environment and sustainable development in regional plans.

The Environment Agency was created in 1996. It took over all the powers and duties of the National Rivers Authority, the waste regulations function of the county councils and air quality functions from Her Majesty's Inspectorate of Pollution. It has a wide range of duties and powers relating to environmental management and improvement in the quality of air, land and water. As far as land use planning is concerned, it is a statutory consultee under the Town and Country Planning Acts. It has the right to be consulted on development plans and on a range of planning applications.

Over the last few years the emphasis has shifted from commenting on planning applications towards influencing development plans. The Agency has also become involved in the preparation of Regional Planning Guidance (RPG) with the result that the current round of RPGs contain comprehensive sets of environmental policies. The advantage is that subsequent development plans can be audited for compliance with their regional policies. This in turn will reduce the number of planning applications which might infringe the Agency's interests. The interface between the Environment Agency and the planning system is set out in the document *Liaison with Local Planning Authorities* (Environment Agency, 1997). The most important sections are set out in Box 9.1:

### Box 9.1: Work of the Environment Agency with the Planning System

Although the Environment Agency operates within an extensive regulatory framework, it must be recognised that its actual controls in respect of development are limited. The Agency is therefore dependent upon effective planning legislation to ensure the protection of the environment and to prevent future problems arising as a result of development.

The Agency liaises with Local Planning Authorities in order to:

- advise on where proposed development may pose a risk to the public or to property from pollution and/or flooding;
- protect the environment from any possible adverse effects of potential development;
- wherever possible, enhance the environment in conjunction with development proposals.

*Regional Planning Guidance*

The Agency will promote our aims and objectives through contact with Planning Conferences and Government Offices at the regional level in the preparation of RPG. We recognise the value of including relevant issues and guidance at all levels of development planning.

*Development Control*

Nationally, the Agency comments on nearly 100,000 planning applications each year. Local Planning Authorities are responsible under town and country planning legislation for consulting the Agency on certain planning applications and have discretionary powers regarding the referral of others.

*Planning Appeals*

The Agency may become involved in an appeal if any one of the following situations arise:

- an objection made by the Agency is included as a reason for refusal;
- an objection made by the Agency was not included as a reason for refusal, but the Agency decides to follow up its objection as a third party, either by a written representation or by appearing at a hearing

> or public inquiry;
> - a recommended condition was objected to by the appellant;
>
> When required the Agency will appear at Examinations in Public and Local Plan Inquiries to support Local Planning Authority policies which accord with our aims and objectives.
>
> *Local Environment Agency Plans (LEAPs)*
>
> LEAPs are non-statutory plans, which will complement areas of environmental policy and regulation for which Local Authorities have executive responsibility. First and foremost LEAPs provide an assessment of the work that the Environment Agency needs to do in a local area.

The Environment Agency has a statutory duty to contribute towards the achievement of sustainable development. One method has been actively to promote good practice in sustainable construction. However, sustainable development is seen as including economic and social issues as well as environmental issues. The Agency has established a reputation of working closely with local planning authorities and the development industry to achieve environmental enhancements through the planning system. It is now seeking ways of working with potential new partners to establish how an enhanced environment can facilitate economic growth and alleviate social exclusion through the regeneration of the run down parts of our towns and cities. This will contribute to the redevelopment of brownfield sites. Sir John Harman (2000), Chairman of the Environment Agency, has summarised the position in the following terms:

- the Agency should be involved in urban regeneration;
- the Agency is concerned about the quality of life;
- poor health and poor environmental quality go together;
- the environment has a key role in solving poverty;
- we have a supporting rather than a leading role in social issues.

Similarly the House of Commons Select Committee on Environment, Transport and the Regions (2000, p. xi) recognised that 'with its remit to promote sustainable development, the Environment Agency is also the body which has the greatest potential for influencing the behaviour and attitudes of Government, business and the general public in such a way that they come to recognise the benefits which are to be gained from acting in a more environmentally sustainably way'. The Agency has therefore needed to extend its planning liaison activities from influencing the formulation of land use plans at regional and local levels, to engaging with a new set of partners in improving environmental quality and reducing poverty and health inequalities. This chapter illustrates the former through an account of regional planning in South East England, before describing the Agency's activities in new partnerships in the Lea Valley in London.

**RPG for the South East**

The current round of RPG has presented the Environment Agency with an unparalleled opportunity to influence the town and country planning system from the top downwards. Furthermore it is a useful channel for promoting sustainable development. The Agency has pursued its objectives through early involvement with regional plan formulation, as well as through formal consultation.

As the preparation of the RPG for the South East is reaching its conclusion, the development of the Spatial Development Strategy for London is just beginning. It is therefore an opportune moment to review how effective the intervention of the Environment Agency in regional planning has been to date and what its prospects may be for the future.

The London and South East Regional Planning Conference (SERPLAN) was asked to provide advice on updating the strategy for the region (RPG 9). In 1996 the Environment Agency was invited to serve on SERPLAN's Natural Resources Group which was charged with advising on a range of environmental and infrastructural matters. At that time it was quite unusual for any body outside local government to be involved directly with SERPLAN.

Over the next two years SERPLAN prepared *A Sustainable Development Strategy for the South East* (SERPLAN, 1998). Its aim was to produce a more sustainable pattern of development and it focused on making better use of the urban areas of the region with a less dispersed pattern of development and a reduced need for travel. Housing provision was to be more closely related to the potential for economic growth. All this would enhance the environment and protect the countryside.

The strategy was based on six policy themes:

Theme 1   Environmental Enhancement and Natural Resource Management
Theme 2   Encouraging economic success
Theme 3   Opportunity and Equity
Theme 4   Regeneration and Renewal
Theme 5   Concentrating Development
Theme 6   Sustainable Transport

The Environment Agency was involved in the policies in *Theme 1* which treats the region's environment as one of its key assets. A high quality environment was seen as essential to the region's future prosperity. The Agency held a series of meetings with SERPLAN. Experts from the Agency's Thames, Anglian and Southern Regions provided advice on water resources, water quality, flood management, waste management, coastal issues, conservation, biodiversity, air quality, contaminated land and sustainable construction.

Much of the input was based on *Thames Environment 21 – The Environment Agency Strategy for Land Use Planning in the Thames Region* (Environment Agency, 1998). The scope and purpose of this initiative in regional environmental planning was set out in the Regional General Manager's Preface:

> Thames Region is already intensively developed and faces pressures for further large-scale development. The Thames Environment 21 strategy provides an approach to achieving sustainable development in these demanding circumstances. It gives the key environmental issues that the Agency wishes to see addressed through the land-use planning system in Thames Region and indicates the enhancement and mitigation measures that are required from developers if the environment is to be protected and enhanced in the way we all want.
>
> Thames Environment 21 will make a significant contribution to regional planning in terms of all the major environmental issues facing the Region. We are particularly keen to discuss these issues on a continuing basis with the Regional Standing Conferences, the London Planning Advisory Committee (LPAC), and the Government Offices for London and the South East. Thames Environment 21 will also assist in providing environmental input to the initial work of the future Greater London Authority and Regional Development Agencies.
>
> We are discussing with the Regional Planning Conference for the South East (SERPLAN), the position concerning water resources. This is one of the most critical environmental challenges facing the Region and could be a constraint on future development (Environment Agency, 1998, p. 3).

Of the eleven environmental policies, those relating to water resources, flood management and the conservation of resources were seen as the most critical for promoting sustainable development in the region. The key points are shown in Box 9.2.

The Environment Agency has already taken the initiative in promoting sustainable construction. In April 1999 it published *Enhancing the Environment – 25 Case studies from the Thames Region* (Environment Agency, 1999a). It has encouraged both local planning authorities and developers to build environmental enhancements into their development proposals. The scope for enhancements includes the following:

- river restoration and channel enhancement;
- surface water run-off attenuation and source control;
- reed bed treatment for surface water;
- conservation, fisheries and landscape enhancements;
- enhanced recreation and education provisions;
- flood plain compensation schemes;
- river bank enhancement works;
- clean-up of contaminated land.

This demonstrates the importance of the Agency not only influencing regional policies in a top-down fashion, but also working to disseminate good practice in construction.

The Agency has also assisted the Thames Valley Economic Partnership – in producing *Quality Living Smarter Housing* (Thames Valley Economic Partnership – to be published). This is designed to show how pressures for housing in the Thames Valley can be achieved in a more sustainable manner and with less environmental impact than is the case with conventional housing. It provides examples of energy efficiency, water efficiency, sustainable urban drainage, as well as landscape and garden design which contribute to energy conservation in buildings and minimal water use for gardens.

The Environment Agency has therefore already made good progress in promoting sustainability into the built environment.

**The Crow Report**

The sustainable development strategy for the South East, published in 1998, was subject to the new form of public examination proposed in draft PPG11. This took place at Canary Wharf in May-June 1999 before a panel chaired by Professor Stephen Crow (Crow and Whittaker, 1999). The Environment Agency was invited to discuss 'The Development and Supply of Infrastructure – Waste and Water'.

**Box 9.2: Advice by the Environment Agency to SERPLAN on the Regional Planning Strategy**

*Water Resources*

The South East is the driest part of the country. The demand/supply balances of the water companies are narrower than in other parts of the country. In certain parts of the region potential shortfalls of supplies to serve the anticipated levels of housing development can be envisaged.

*Flood Management*

Pressures for development could result in new housing being built in areas at risk from flooding. Climate change could raise this risk. The Agency provides local planning authorities with indicative flood plan maps. The Agency is seeking to supplement this basic information with an indication of the relative risk of flooding of specific sites, and how such a risk might be managed.

*Sustainable Construction*

The policy on Conservation of Resources itemises aspects of sustainable design, which should be provided through development plans, design guides and good practice notes.

The design of development will be a major influence on the extent to which new development is sustainable, and will cumulatively be regionally significant. Aspects of sustainable design include:

- use of waste prevention and minimisation techniques or failing that the installation of pollution abatement technology to reduce emissions to air and water;
- control measures for surface water drainage as close to its source as possible (including the attenuation of runoff to prevent flooding or erosion of watercourses);
- building designs which facilitate the use of renewable energy;
- the use of combined heat and power (CHP);
- energy efficient installations, including passive solar design for buildings and improved insulation;
- water efficient installations, including the use of grey water systems; the use of renewable and recycled materials during construction; and
- design to facilitate recycling systems, including energy from waste.

A crucial aspect of debate was related to the level of housing proposed in the period up to 2016 (see Table 9.1). The issue was whether SERPLAN's capacity-based approach, involving 'Plan, Monitor and Manage', was the correct one or whether the full demographic projections based on the Government's 'Projections of Households in England to 2016' should prevail. 'Plan, Monitor and Manage' was derided as 'fudge, dither and panic', and the Panel recommended the higher household projections as a basis for housing provision. However, draft RPG 9 (Government Office for the South East, 2000a) has since indicated that the Government has abandoned the 'predict and provide' approach of Professor Crow, and that it would support 'Plan, Monitor and Manage'.

The Panel considered water resources, water quality and flood management in the light of the higher figure. It took the view that water supply was but one factor influencing the location of new development. 'Social and economic considerations may often be more important and if they are then it is simply up to the water companies to do their duty cheerfully.'

The Panel did nevertheless recommend that RPG should 'encourage active conservation of water... to ensure that all new developments meet these high standards of water efficiency' (Crow and Whittaker, 2000).

There was general agreement that development posed an unacceptable risk to the quality of water in the region's rivers. The Panel recognised that water quality was a material factor in considering future development allocations.

The reaction to the 'Crow Report' was split between the development industry, which was surprised and pleased at the recommendations for greater levels of housing, and SERPLAN and the local planning authorities who faced problems in finding sites for the additional development. The Environment Agency registered its concerns with the Government Office of the South East (GOSE) that a greater amount of housing aggravated its problems in terms of water supply and quality, and perhaps also increased the number of people who might be at risk from flooding, should additional development be permitted in the region's floodplains.

**Table 9.1: Proposed Annual Housing Rates for South East England**

|  | *PROPOSED ANNUAL HOUSING RATES* | | |
|---|---|---|---|
|  | **LONDON** | **REST OF SOUTH EAST** | **TOTAL** |
| RPG 9 1994 Additional Dwellings 1991–2006 | 17,333 | 39,667 | 57,000 |
| 1992 Household Projects 1991–2016 | 23,000 | 44,120 | 67,120 |
| SERP500 Dec. 1998 Baseline | 22,000 | 34,480 | 56,480 |
| b) Indicative Range to be planned for | 22,000<br>22,000 | 35,680<br>36,560 | 57,680<br>58,560 |
| Report of Panel (Crow Report) | 22,000 | 54,925 | 76,925 |
| Draft RPG (March 2000) | 23,000 | 43,000 | 66,000 |
| Draft and Revised RPG (Dec. 2000) | 23,000 | 39,000 | 62,000 |

**The Draft RPG**

Draft Regional Planning Guidance in the South East, RPG 9, (Government Office for the South East, 2000a) was published in March 2000 to general criticism all round. The housing figures for the 'Rest of the South East' (ROSE), which excludes London, were a significant reduction on the recommendation of the Crow Report but still well in excess of SERPLAN's recommendation. The guidance, however, sought to minimise the impact on the environment by encouraging higher densities in urban areas. The public expressed concern about the potential impact of development on the high quality environment in the South East.

SERPLAN not only challenged virtually every policy but also decided to recommend their original housing figures, as these had a coherent robust basis. Furthermore they expressed concern that their theme-based approach had been abandoned.

The Environment Agency raised many detailed criticisms and also stressed that the RPG could result in unacceptable gradual erosion of the environment. It also objected to the failure of the RPG adequately to promote sustainable development. In particular water resources were not treated as a constraint on development and the policy on water quality had been omitted altogether.

In a press release accompanying its response the Agency said: '... it is worried that proposed levels of growth in the region will result in unacceptable detriment to the environment unless the pressures they produce are managed to achieve a sustainable outcome. The Agency is particularly concerned that there are insufficient water resources in areas such as the Thames Basin and Ashford in Kent to support planned development. It has also pinpointed towns such as Basingstoke, Ashford and Aylesbury where river water quality may be downgraded as sewage treatment works struggle to cope with increased volumes of effluent' (Environment Agency, 2000).

In December 2000 *Draft Revised Regional Planning Guidance for the South East* was published (Government Office for the South East, 2000b). The policy content was in its final form. However, it contained a further reduction of the housing target for 'ROSE' and initiated a further consultation exercise on the distribution of houses amongst the counties. An accompanying letter from the Minister of Housing and Planning indicated that the revised lower figure would only last until 2006. A higher rate of provision would be likely to be necessary from 2006 onwards to meet long-term needs. Such an approach, it may be argued, is not a satisfactory situation for County Councils who need to roll forward their structure plans for 15 years. The resulting uncertainty will be a severe test for the 'plan, monitor and manage' approach to housing provision in the region.

**The Spatial Development Strategy for London**

The Greater London Authority (GLA) took office in July 2000. The Environment Agency had been working for a year before this with the shadow organisation to ensure that the environment is treated as a critical part of the city's natural capital. The Agency monitored the GLA Bill through parliament to make sure that the Environment Agency would be consulted on the various strategies that the GLA will have to prepare. We also held conferences and breakfast meetings to identify the environmental issues that are of most concern to Londoners (Environment Agency, 1999b). Several 'cross-cutting' themes emerged as being of prime concern to Londoners. These were:

- London as a world class city;
- the health of its citizens and the influence of environmental conditions on their well being;
- the need to foster environmental awareness and civic pride;
- the role that ecological networks could play in urban regeneration;
- concerns about capacities of sites and the potential density of development taking place on them.

The Agency is now engaged with the GLA in seeing how these issues may be developed and given expression in the GLA's strategies. Work is at an early stage but it is possible to give an indication of the potential that these cross-cutting themes may open up. This chapter looks at two of these issues at the scale of London as a whole, then illustrates how the Agency is developing these new initiatives in the Lea Valley.

*The Health of London's Citizens and the Influence of Environmental Conditions upon their Well Being*

Sustainable development is seen as including economic and social issues as well as the environment. London is a world-class city: a world-class financial centre, a world cultural centre, and a world heritage centre. Nevertheless it contains some of the worst concentrations of urban deprivation and unemployment in the country in terms of both extent and intensity. Significant swathes across London suffer from declining industries, social exclusion and degraded environments. Parts of inner London including the Lea Valley are characterised by above average unemployment rates, high levels of social deprivation, low skills levels, low educational attainment, poverty, poor health prospects, dependence on declining industries and derelict urban fabric.

Urban regeneration has been built on linking economic issues with land use planning. However, social issues in general and social exclusion in particular have not so far been well integrated into either economic or land use planning. Regeneration has been most successful where local authorities have moved away from their role of service providers and have acted as a catalyst for creating partnerships. The new Greater London Assembly has limited direct powers and its success will depend on how far it is able to develop this catalytic role.

The Government is committed to reducing health inequality (Shaw *et al.*, 1999). However, the gap between the health of rich and poor is widening. The Government's thinking appears to be less than wholly joined up. However, the Social Exclusion Unit looking at the relationships between poverty, health and regeneration does recognise the effects of environmental conditions (Cabinet Office, 2000).

*Planning Policies*

A number of areas of regional significance are identified in the Draft Regional Planning Guidance for the South East as Priority Areas for Environmental Regeneration (PAERs). The criteria for designation include above-average unemployment rates, high levels of social deprivation, low skills levels, dependence on declining industries, derelict urban fabric, peripherality and insularity. These areas need tailored regeneration strategies, backed up by appropriate resources, to address their problems and maximise their contribution to the social and economic well being of the region.

The Draft Regional Planning Guidance seeks to adopt an all-embracing approach to these issues by including the following policies:

- the quality of life in urban areas, including suburban areas, should be raised through significant improvement to the urban environment, making urban areas more attractive places in which to live, work, shop, spend leisure time and invest, thus helping to counter trends to more dispersed patterns of residence and travel (Q2);
- health, education and other social infrastructure requirements need to be taken into account fully in development planning throughout the Region (Q6);
- in order to address strategic spatial inequalities around the region, particular attention should also be given to actively supporting economic regeneration and renewal in Priority Areas for Economic Regeneration (PAERs) (RE7).

Similarly PPG3 states: 'The Government intends that everyone should have the opportunities of a decent home. They further intend that there should be greater choice of housing and that housing should not reinforce social distinctions. New housing and residential environments should be well designed and should make a significant contribution to providing urban renaissance and improving the quality of life' (Department of the Environment, Transport and the Regions, 2000).

It is clear that the planning policies are comprehensive in their approach to social and health issues but that mechanisms implementing these policies have some way to go. Tackling these issues is going to involve creating a range of new partnerships. The Environment Agency is used to working closely with local planning authorities, the development industry, the water companies and environmental interest groups. To deal with overlapping social and environmental issues will mean involving bodies such as the Department of Health, the British Medical Association, health trusts, Directors of Public Health and Education and the voluntary sector. The catalytic role of the GLA should encourage these linkages to be forged.

*The Role that Ecological Networks could play in Urban Regeneration*

London's waterways, including the Thames tributaries, substantially coincide with the city's Metropolitan Open Lands. A high proportion of Sites of Special Scientific Interest lies in these areas. The river corridors therefore offer a network for nature conservation, biodiversity and recreation. There will be an emphasis on restoring missing links in the network. The completion of the Thames path through the city provides an excellent example of what can be achieved in this respect. A recent report has indicated that although the quality of river water has improved across the country as a whole, there are significant cases in London where river quality is poor. This is caused by combined sewer overflows and plumbing of foul sewage into surface water drains. Water quality is an important element in urban regeneration. The development of waterside sites in, for example Leeds (Howes, 2000a), Newcastle and Salford could not have taken place without the prior improvements in water quality.

Many of the locations for future major developments are adjacent to London's waterways. Whether they will be treated as a liability or an asset will depend on whether sufficient improvements in water quality can be achieved. Integrating phased improvements in water quality into the town and country planning system is a challenge that the Environment Agency is trying to address.

The town and country planning system has traditionally dealt with spatial issues. It has relied on constraints that can be readily shown on a map as its basis for environmental protection. Increased public pressure for addressing a wide range of environmental issues is resulting in an integration of urban physical planning and environmental management.

The River Ecosystem scheme, whilst a useful measure of water quality for the purposes of the Environment Agency, does not readily translate into land use planning. A better approach would be to recognise that incremental improvements to water quality do open up additional opportunities for land uses and development in general and for activities related to the waterside in particular.

Table 9.2 shows how local planning authorities could set specific objectives to be achieved during the lifetime of their development plans. The next step will be draft policies for local planning authorities which will give practical effect to this high level aspiration.

A local planning authority could for example aim to move from Stage 2 to 3 within the first five years of a development plan and then from Stage 3 to 4 in the second five years. This would link into a phased programme of urban regeneration.

## The Lea Valley

In order to examine the significance of the overlap between these social and environmental issues more fully, the Agency has been studying a specific part of London, which is characterised both by social exclusion and by relatively poor environmental conditions.

The Lower Lea Valley (from the M25 to the Thames) presents a major opportunity for improving economic, social and environmental conditions (Lee Valley Regional Park Authority, 1998). Significant incidence of deprivation and social exclusion is matched by poor water quality, air quality problems and concentrations of contaminated land.

Until the 1970s the area was a significant manufacturing centre and relatively prosperous. The Lea Valley was home to a profusion of important industries including the design and manufacture of ships, boats, explosives, armaments, porcelain, bricks, perfume, chemicals, plastics, furniture, floor and wall coverings, vehicles and their accessories, rubber commodities, footwear, clothes, alcoholic beverages, musical instruments, office equipment, electronic and electrical goods. Many well-known companies disappeared from the area. These include Bryant & Mays, Ediswan Lamps, Belling, Thermos, Thorn EMI, Royal Small Arms, JAP, Lebus Furniture and Gestetner (Lewis, 1999).

**Table 9.2: Changes in Water Quality and Attractiveness of Area for Developers**

| Threshold | Outcome |
|---|---|
| 1) Very poor water quality | Developers turn their backs on the river and it has no recreational appeal. |
| 2) Activities of volunteer groups in clearing rubbish and planting reeds, stabilising banks and encouraging community projects | A less intimidating environment encourages access and a presumption against depositing litter and waste. |
| 3) Upgrading of water quality so that it does not smell | Developers are prepared to face the river rather than back onto it. |
| 4) Upgrading of water quality to a basic fishery standard | Wildlife returns, as do anglers and basic forms of waterside recreation. A waterside location becomes a significant selling point for properties. |
| 5) Upgrading to good quality fisheries | Developers are prepared to make water a major feature of development. Contact water sports become possible and restaurants, bars and cafés are attracted to the waterside. |

The loss of Lea Valley industries in recent years has been caused by shifts in the manufacturing base. A rapid decline began in the 1970s, which resulted in a downturn in the local economy and social deprivation, with an urgent need for regeneration through public/private partnership.

*Regeneration*

The whole area enjoys European Union (EU) Objective 2 status. This means that European funds can be made available for projects, which will initiate the social and economic regeneration of the area on a basis of matched funding. All six London Boroughs in the Lea Valley have submitted bids under the various Single Regeneration Budget 'rounds'. Single Regeneration Budget funding enables substantial private sector investment to be levered into the area for the benefit of the local community.

Most of the Single Regeneration Budget bids reflect the fact that parts of the Lea Valley are amongst the most deprived in the country. The baseline information for one typical submission is shown in Box 9.3.

In recent years a 'halo' effect from regeneration in Docklands and Stratford has taken place. The opening of the Jubilee Line extension has enhanced this. House prices in the London Boroughs of Newham and Tower Hamlets are rising rapidly, which may prove to be a spur for regeneration. However this has not prevented the proximity of deprivation and poverty in, for example, the Crossways estate in Tower Hamlets next to the conspicuous wealth of Canary Wharf.

Priority is given to tackling the issues set out above. Environmental measures often consist of nothing more than improvements to the housing stock. Nevertheless there is substantial scope for improving the underlying situation relating to water quality, air quality and contaminated land in the valley as a whole.

---

**Box 9.3: Deprivation Indicators in a Single Regeneration Budget Bid in the Lea Valley Area, London**

- a registered unemployment rate of 16% and 28% for black and ethnic minorities;
- a mean weekly income per household of £169.80;
- the standardised mortality rate for men aged 15–64 is 76% higher than the national average and for women 50% higher;
- illness rates are 40% higher than the national average;
- no direct public transport route to a general hospital providing outpatients and Accident & Emergency services;
- the number of recorded crimes and offences per 1000 people is more than twice the figure for the Borough and two and a half times the national average;
- 58% of people feel unsafe at night when walking alone;
- poor housing is the greatest source of local dissatisfaction (37%);
- a tenure mix where only 11% are owner-occupiers or leaseholders (72% are council tenants);
- Key Stage 2 and GCSE results improving but in a number of instances considerably below the national average.

*Water Quality*

A fundamental problem for water quality in the River Lea is the shallow gradient and consequent low velocity leading to deoxygenation. The multiple sources of pollution of the River Lea demonstrate how complex it is to achieve significant improvements in water quality. These sources are: effluent from sewage treatment works, stagnation, surface water run-off, combined sewer overflows, misconnection of domestic plumbing, discharge of untreated sewage at times of heavy flow, and saline incursions at times of high incoming tides when salt water can sometimes flow over the top of Bow Locks.

The Agency is involved in the Lea Valley Partnership, which is running a project, which includes river restoration, habitat provision and bankside improvements. They have a Single Regeneration Budget of £5m for a seven-year programme to make a fundamental difference to water quality in the lower Lea. However, the quantity of surface water run-off from a largely urban catchment will prevent further improvements in water quality, as will the high bacteriological content which will pose a risk to health. The Agency is therefore trying to integrate its support for physical improvements with the social and economic strategy for the area.

*Land Contamination*

The Government's policy is to encourage the redevelopment of brownfield sites. The Agency wishes to encourage the use of such sites as it can provide a positive contribution towards sustainable development with environmental, economic and social benefit, and the prudent use of a finite resource.

There are extensive old landfills, some of which are unlicensed. Local planning authorities have to publish a contaminated land strategy by July 2001. Until then, it is illegal to identify unlicensed sites in public. These are a serious issue in the Lea Valley. The proposed after-use of such sites determines the level of remediation. In assessing the proposed remediation, the Agency's comments as a consultee in the planning process are limited to the impact of the work on the aspects of environment for which the Agency has responsibility. Issues of risk to future inhabitants rest with the Environmental Health Department of the Local Authority.

Nevertheless, the Agency, through its work with a wider range of partners, is developing new networks to deliver a more rounded form of sustainable development. Advice is now available (SNIFFER, 1999), to ensure that appropriate action is taken to deal with existing contamination where it poses unacceptable risks to human health or to the environment. It stresses that even where expert assessment may indicate tolerable levels of risk, community perceptions may be very different.

*The Lee Valley Park Authority*

Such community perceptions of risk are evident in a scheme currently being assessed by the Lee Valley Park Authority. The Authority was set up in 1967. Since then, it has concentrated on transforming many areas of neglected desolation. It has acted as a direct provider of services. The emphasis is now to shift to developing the park and its open spaces for informal leisure and recreation. It will also promote formal sports and leisure. Its role will shift to that of leader, enabler and facilitator in line with the new philosophy for local government of 'Best Value' (Lee Valley Park Authority, 1998).

Plans are being formulated to build a stadium for the World Championship Athletics in 2005. It will be on the site of the Pickett's Lock Leisure Centre. At the time of writing, the Lee Valley Park Authority is carrying out feasibility studies. A planning application was expected in April 2001. Concern has been expressed about the construction of an Olympic standard athletics stadium adjacent to the Edmonton Incinerator, which is about to be substantially expanded. However, the potential impact of the emissions would be insignificant against a background of vehicle emissions. The flue from the new incinerator will result in faster, hotter emissions, which will rise further before spreading out. The effect will be to reduce levels of air pollution in the immediate surroundings.

**Conclusions**

Active participation in regional planning has been a novelty for the Environment Agency and the benefits are becoming apparent. There are significant contrasts between our experiences with RPG for the South East and the Spatial Development Strategy for London.

The Environment Agency's involvement with RPG 9 was essentially concerned with the Agency's functions. Originally SERPLAN sought advice on water resources and flood risk, but the scope of our advice was subsequently substantially widened. The GLA has initially sought advice on water issues. However, our input is likely to develop on the lines of the cross-cutting themes set out above.

The RPG process has been lengthy – more than four years to date. In contrast the GLA is required to produce its strategies to a very short timetable. Whereas SERPLAN's strategy was comprehensively thought through the GLA is bound to restrict itself to identifying issues with back-up material from earlier studies by the London Planning Advisory Committee.

RPG will provide advice for local planning authorities to prepare their development plans. Should the advice be ignored and policies on key issues omitted, there is a sound basis for challenging the content of development plans – a valuable additional safeguard for the Agency. However, bolstering Agency objectives by development plan policies has already proved very helpful. The Agency has, for example, found that where there is no policy which protects the floodplain from development, its advice on individual applications for development in the flood plain has not always been heeded. However, where there is such a development plan policy, this advice is only ignored under very rare circumstances. The GLA's strategies, whilst they will provide some advice to London Boroughs, will also seek to influence a range of issues which are normally outside the scope of land use planning.

There is little doubt that the Environment Agency has improved its effectiveness in the town and country planning system through its involvement in regional planning. This applies not only to its functions but also through policies covering sustainable construction, which to date has been the most obvious manifestation of making a contribution towards sustainable development.

## Note

[1] The author is grateful to the Environment Agency for permission to contribute to this chapter. The views are those of the author and do not necessarily reflect the policies of the Environment Agency.

## References

Cabinet Office (2000), *National Strategy for Neighbourhood Renewal – A Framework for Consultation*, Social Exclusion Unit, Cabinet Office.

Crow, S. and Whittaker, R. (2000), *Regional Planning Guidance for the South East of England: Public Examination May-June 1999 – Report of the Panel*, GOSE, Guildford.

Department of Environment, Transport and the Regions (2000), *Planning Policy Guidance Note 3*, DETR, London.

Environment Agency (1997), *Liaison with Local Planning Authorities*, Environment Agency, Bristol.

Environment Agency (1998), *Thames Environment 21, The Environment Agency Strategy for Land Use Planning in the Thames Region*, Environment Agency, Reading.

Environment Agency (1999a), *Enhancing the Environment – 25 case studies from the Thames Region*, Environment Agency, Reading.

Environment Agency (1999b), *A Greenprint for London*, Environment Agency.

Government Office for the South East (2000a), *Draft Regional Planning Guidance for the*

*South East*, GOSE, Guildford.
Government Office for the South East (2000b), *Draft Revised Regional Planning Guidance for the South East*, GOSE, Guildford.
Harman, J. (2000), *Speech to Urban Lifestyles Conference of Institute of Landscape Architects, Newcastle*, Environment Agency.
House of Commons Select Committee on Environment, Transport and the Regions (2000), *The Environment Agency: Sixth Report*, HMSO, London.
Howes, H. (2000a), 'Urban regeneration, the water element', *Water and Environment Manager*, vol. 5 (4).
Howes, H. (2000b), *Towards a Healthier Environment for London*, Urban Lifestyles, Balkeman.
Lee Valley Regional Park Authority (1998), *Strategic Business Plan 2000-2010, A Fresh Direction*.
Lewis, J. (1999), *London's Lea Valley*, Phillimore, Chichester.
SERPLAN (1998), *A Sustainable Development Strategy for the South East*, SERP 500, SERPLAN, London.
Shaw, M., Darling, D., Gordon, D. and Dancy Smith, G. (1999), *The Widening Gap*, Policy Press, Bristol.
SNIFFER, (Scotland, Northern Ireland Forum for Environmental Research) (1999), *Communicating Understanding of Contaminated Land Risks*.
Thames Valley Economic Partnership (to be published), *Quality Living Smarter Housing*.

# 10 Regional Planning and the Environmental Dimension of Sustainable Development

PETER ROBERTS

**Introduction**

The intentions of this chapter are threefold. Firstly, to examine the evolution of the environmental dimension of regional planning in Britain, and especially in the English regions. Secondly, to assess the extent to which the requirements of the environmental agenda have been acknowledged and incorporated into the re-defined theory and practice of regional planning and regional development (the two are now inseparable). Thirdly, to speculate on the possible role of the environmental dimension of sustainable development as a key element in the evolution of regional planning and development over the next decade.

In presenting this review and analysis it should be recognised that an important consideration for policy makers and planners alike, is how to achieve environmentally sound regional planning and development without introducing the need for additional expenditure and without adding extra layers of administration or bureaucracy. Indeed, one of the objectives of a programme of reform should be the promotion of the most effective use of all resources. A further distinguishing feature of the current debate is that the technical-professional processes of regional planning and development are now inextricably intertwined with the new political-cultural paradigm, which itself is derived from subsidiarity, devolution and a desire to introduce a higher level of regional self-determination and co-ordination into both policy-making and implementation.

Two of the points made in the preceding paragraphs are of considerable importance and have resulted in the transformation of regional planning practice in the United Kingdom (UK) during recent years. The first of these is that regional planning and regional development, which for most of the period since 1945 have diverged (Baker, Deas and Wong, 1999;

Roberts and Lloyd, 1999), have recently emerged as complementary activities that share a number of common features and also share a focus on the delivery of sustainable development. As will be seen later, the traditional scope and competence of the total set of regional planning and development activities has widened considerably, whilst the boundaries of individual elements of this set of activities have become less rigid or distinct. One of the major consequences of this reformulation of regional planning and development functions is that different policy 'processes' and 'products' have emerged, some of which are shared between two or more regional planning and development activities. Regional planning and development has followed the model of partnership and collaboration pioneered in the urban policy arena (Chapman, 1998; Healey, 1997), and now forms part of the wider English regional project. Similar developments can be observed in Wales and Scotland (see the chapters by Alden and Lloyd in this volume). A second factor that has contributed to the evolution of this more comprehensive and co-ordinated approach is associated with the realisation that sustainable development cannot easily be imposed from above; the reality is that action from above and below are both required, and that the region provides an intermediate level at which strategies and programmes of implementation can be negotiated, agreed, monitored and reviewed. As Possiel *et al.*, have argued 'participants in the regional planning exercise, though representing individual interests, have a shared commitment to rules and procedures of the process that can be controlled' (Possiel *et al.*, 1995, p. 12). Regional planning and development can, therefore, now be considered as part of the wider territorial governance which has recently emerged in the UK (Roberts, 2000).

Following a brief exploration of some of the key events and lessons from the history of attempts to incorporate environmental concerns in regional planning, this chapter moves to a consideration of some of the major issues under discussion in the current debate on the reformulation of regional planning and development. A major factor in this debate is the need to consider the extent to which regional policies and governance are essential elements in the achievement of the goal of sustainable development. The final section of the chapter looks forward to the future development of environmentally sustainable regional planning.

## A Brief History

The origins of regional planning, and especially the environmental dimension of regional planning, can be traced back to contributions made by various academics and practitioners during the late nineteenth century. These theorists and social activists were concerned with a number of problems which are still familiar, including those associated with the negative consequences of rapid industrial growth and the presence of poor urban living conditions, and some of them were especially interested in the consequences of the divorce of people and economic activities from nature – what we now call the environment (Cronon, 1991). Social, physical and environmental degradation were frequently the outcome of unthinking industrialisation and urbanisation, and the results of these processes of change were evident across large tracts of the UK. Ebenezer Howard visited many urban regions in Europe and the United States before returning to England to write 'Tomorrow: A Peaceful Path to Real Reform', which was first published in 1898 and republished in 1902 as 'Garden Cities of Tomorrow'. An often forgotten, neglected and misunderstood element of Howard's work, is that his notion was a regional one and that he was advocating not a model village on a grand scale, but a balanced and environmentally sound 'social' city-region that could grow in a planned manner through the development of a linked series of polycentric settlements (Hall, 1974).

Howard's ideas are important because they provided a practical solution to the problems of rapid urbanisation which were experienced during the nineteenth and early twentieth centuries. The key features of relevance today include: the emphasis that he placed on social reform and justice, the need for the effective management of the environment, the achievement of economic progress within the limits of the two preceding characteristics, the need to think and plan at a regional or sub-regional scale, and a particular concern to ensure the provision of an effective link between spatial planning and the various processes of territorial management.

If Howard was a practical idealist, the ideas of the second 'founding father' of environmentally sound regional planning can best be described as those of a realistic visionary. Patrick Geddes, drawing on the work of French and German human geographers, argued in 'Cities in Evolution' (1915) that because industrial and urban growth affected not only the town but also the surrounding hinterland, and because this growth exerted considerable pressure on environmental resources across an even wider area (what we now call the 'ecological footprint' of a city-region), then planning should be undertaken at a regional scale and that this planning should be chiefly concerned with the effective treatment of the failures of what Geddes called the 'Paleotechnic order' of unthinking and uncaring urbanisation and industrialisation. It is, perhaps, a source of discomfort that many of the weaknesses of planning that were identified by Geddes – what he described as dissipating resources and energies, depressing life, unemployment, disease and folly, indolence and crime, among others – still persist, and that it is only now that we fully understand and appreciate his call to deal with these problems through the adoption of an integrated and co-ordinated approach to policy design and implementation.

Other key figures from the history of regional planning and development who were influential in stimulating the debate on the case for a 'balanced' approach to planning, include Mumford (he was instrumental in establishing the Regional Planning Association of America and transmitted the ideas of Geddes to a wide audience of academics and practitioners), MacKaye, who provided much of the initial impetus for the practical application of the work of Geddes in the USA, Abercrombie, who sought to distil the essence of the thinking of Howard, Geddes and others and apply it to the balanced planning of the great industrial city-regions of the UK and elsewhere, and Unwin, who developed Howard's ideas and attempted to apply them, often in resource-constrained situations, to provide quality urban environments for the accommodation of mass residential development.

Although many of the ideas, principles and distinguishing characteristics of this late nineteenth and early twentieth century flowering of regional planning thinking and practice incorporated what we now refer to as the environment as a central element, in many other situations planning practice totally ignored or, in some cases, abandoned the 'balanced' approach. Nevertheless, both in the UK and elsewhere, and arguably especially in the United States, the origins of the current attempts to promote environmentally sustainable regional planning can clearly be identified in the practice of the past (Friedmann and Weaver, 1974; Roberts, 1994). From this early period we can identify the key tasks that were, and still are, associated with the promotion of 'balanced' regional planning and development:

- a concern with the conservation of natural resources;
- an attempt to control and manage the flow of commodities;

- an assessment of the costs associated with the use of resources and the assignment of these costs to an appropriate person or organisation;
- the promotion of social welfare, justice and health;
- the support of economic progress, especially through an approach that respects the capacity of the environment; and, most significantly;
- the promotion of 'territorial integration' as a means of planning and managing the economy-society-environment relationship; a relationship which we now refer to as sustainable development.

This brief glimpse of history allows us to state, with some degree of confidence, that modern day regional planners have not 'discovered' sustainable regional planning. Rather, regional planners are only just beginning to 'rediscover', or 'recover', the lost art and science of 'balanced' planning and development. This 'rediscovery' has helped to establish the parameters of the current debate and to set the style and scope of sustainable development policy at all spatial levels. As part of this process of 'rediscovery' it is acknowledged that the region – at least in terms of its meaning as a 'natural' region – offers an ideal spatial vehicle for the purposes of planning and development that is in accord with the objectives of environmental sustainability (Cohen, 1993). As Gibbs has observed, 'there is an increasing awareness that environmental protection and regional development should be complementary activities' (Gibbs, 1998, p. 365). The reason for this coincidence between regional planning and development, on the one hand, and environmental requirements, on the other hand, has much to do with the scale and composition of regions, especially 'natural' regions, many of which are river basins, and with the fact that many sustainable development challenges are presented 'particularly clearly at regional level' (UK Round Table, 2000, p. 1). The region, in summary, is of a scale that is large enough to allow for effective strategic planning and implementation, but small enough to relate to the individual elements or characteristics that are evident at local level within the defined territory. As such, the region can be seen to provide a useful intermediate or 'meso' level for the purpose of planning and development. This question of scale – what can be called 'spatial fit' – is particularly important with regard to environmental issues. After all, the classic Geddesian 'valley section' contained most of the elements necessary for urban living within the confines of a single region.

## Reformulating Regional Planning in England since the 1970s

Although the preceding section of this chapter has suggested that regional planning in the UK 'lost its way' from the 1930s onwards, this is not entirely correct. Whilst many of the regional plans and strategies that were prepared in the English regions from the late 1940s to the early 1990s were principally concerned with the promotion and accommodation of economic growth, there were a number of notable attempts to provide a more 'balanced' approach. This emergence of a more 'balanced' approach can be seen in some of the early regional strategies – the warning bells were sounded by the Strategic Plan for the North West (1974), which emphasised the need for environmental and social problems to be attended to in order to ensure that they would not constrain or damage the future development of the regional economy – and these strategies began to move towards the adoption of an approach which encapsulated some of the features of the earlier 'balanced' model. Other warnings of the problems for a regional economy that could be caused by a neglect of environmental conditions were contained in various official reports, including the Hunt Report on the Intermediate Areas (1969), which placed particular emphasis on the environmental, economic, visual and social damage associated with the presence of derelict land.

However, despite the growing frequency and severity of environmentally related protests and anxieties that emerged during the 1970s and 1980s, the real 'rediscovery' of the need for balanced regional planning and development did not occur until the late 1980s. As Breheny (1991) has observed, business, environmental groups and local NIMBYS, especially in the South East of England, helped to rehabilitate regional planning through their calls for clearer and more detailed policy and strategy, especially in relation to the planning and provision of major infrastructure. These 'bottom up' pressures for the introduction of a more comprehensive and sustainable approach to regional planning and development, were reinforced and given official backing by the publication of the report of the World Commission on Environment and Development, the Brundtland Report, in 1987. This report proved to be hugely influential in changing the attitudes and policies of governments at all levels, including those of the member states of the European Union (EU).

In the UK, the influence of the Brundtland Report (as reinforced and extended through the agreements reached at the Rio Summit in 1992 and elsewhere), together with a growing policy awareness and concern expressed by the European Union, influenced the introduction of a series of improvements in the environmental dimension of planning and development policy, including the enhancement of central government policy guidance on the environmental content of regional plans and development programmes. This process, whereby sustainable development concerns were gradually incorporated into regional planning, was reflected in the increased emphasis placed on environmental matters in Regional Planning Guidance (RPG) (Baker, 1998; Roberts, 1996) and in the introduction in some regions of a more proactive approach to the planning and management of environmental issues at a regional scale (Carter and Winterflood, 2000) which went beyond the RPG process. Equally, Planning Policy Guidance (PPG) issued by central government (under variously numbered PPGs, 12, 15, 12 and now 11) gradually resulted in an increased emphasis on the environmental component of regional and other plans.

The various outcomes of this process of 'rediscovery' are evident in the present situation in which RPG and a number of associated regional planning and development exercises are now orientated towards the achievement of sustainable development. This new orientation and positioning of the various elements of English regional policy, together with the enhancement of the environmental component of a range of associated areas of activity, such as the regional programmes or Single Programming Documents (SPDs) prepared with support from the EU's Structural Funds (European Parliament, 1998) and the gradual introduction of more rigorous and comprehensive forms of environmental impact assessment (Glasson, 1995), has resulted in the acceptance of the achievement of sustainable development as one of the primary objectives of virtually all regional planning and development exercises.

Examining this somewhat unexpected recent rise of environmentally sustainable regional planning and development, the most striking feature is the general coincidence of intent which now exists between the various aspects and strands of environmental spatial intervention policy. For example, the internal co-ordination of spatial policy within the institutions of the EU (especially between the directorates concerned with the Structural Funds [XVI], Environment [XI] and Research and Development [XII]) has coincided with the re-definition of the role, content and operation of RPG (DETR, 2000a), the provision of guidance to the Regional Development Agencies (RDAs) on the formulation of their strategies (DETR, 1999) and the more recent introduction of a requirement to prepare Regional Sustainable Development Frameworks (RSDFs) in the

English regions (DETR, 2000b). Whilst this coincidence of policy is the result of both the 'cascade' of priorities from Europe and an attempt to produce greater policy coherence between the various strands of central government activity in the English regions, it is surprising, on the basis of previous performance, how rapidly the desired coincidence has emerged in concrete form.

In addition to the elements of regional planning and development that have been mentioned so far – RPG, RDA strategies, RSDF and the regional programmes supported by the EU Structural Funds – a number of other regional strategy exercises have been introduced during the past three years, some of which play an important role in the negotiation and implementation of regional sustainable development agendas. Examples of such exercises include the Environment Agency's Local Environment Agency Plans (LEAPs) which set out integrated plans for river basin management, regional minerals plans, biodiversity action plans at regional and local levels, regional waste plans, regional forestry strategies, and regional transport strategies, which are now incorporated into RPG (Ash, 1999).

**Mechanisms for Regional Integration**

An important new consideration in the English regions is how best to co-ordinate and integrate the growing number of regional and sub-regional strategies, plans and programmes. There are two important issues here. Firstly, there is a danger that the process of co-ordination will simply conflate poor plans and strategies, thereby producing an integrated strategy which reflects all of the worst features of the component plans. Secondly, the co-ordination of plans in the absence of an integrated programme of implementation will simply result in regional partners paying lip service to the need for co-ordination, whilst still pursuing separate objectives through independent expenditure processes. In the absence of an elected regional authority that can exert both direct and indirect influence over the activities of regional bodies, it has been argued that there is 'obvious potential for conflict between local authorities and business-led RDAs' (Tomaney, 1999, p. 78) or for disagreement between a voluntary regional chamber/assembly and the Government Office (GO) in a region. This suggests that the introduction of democratic control and accountability over the full range of regional planning and development activities may well be a pre-condition for the introduction of an effective approach to sustainable development.

All of the above points can be considered to be a reflection of the mode of evolution which has typified the formulation of the regional agenda in the English regions. A piecemeal process of development, albeit one which has accelerated during the past three years, has resulted in the establishment of a set of parallel regional initiatives, many of which are still guided from the centre. This process has taken place in the absence of a core or overarching regional vision and strategy, established at the outset, which can be used to guide all individual topic or subject strategies and thereby ensure that they follow a common sustainable development agenda and that they express their contributions in accord with an agreed spatial framework. In part the absence of a 'leading' regional strategy reflects the institutional fragmentation which has dominated the history of regional planning in England (Baker, Deas and Wong, 1999; Roberts and Lloyd, 1999), but it can also be seen as a manifestation of the internal battles within Westminster and Whitehall for control of the regional agenda (Murdoch and Tewdwr-Jones, 1999). Belatedly, and somewhat ironically, a requirement to produce RSDFs has been introduced. These documents, which will be published by the end of 2000 – a year or more after the production of SPDs, RDA strategies and many RPG final documents – are required to 'identify the links between different regional activities and initiatives and set a common and agreed high level vision for sustainable development at regional level' (DETR, 2000b, p. 7). What is in doubt is the ability of RSDFs to 'retrofit' a common sustainable development strategy on existing regional plans and strategies. It is more likely that RSDFs will be used to guide the review and roll forward of other strategies.

A further matter that requires attention is that although RPG is a key element in the new regional arrangements in most regions, it is RSDF that is most likely to provide an overall context for the development of sustainable development policy. As can be seen in a number of regions, what has occurred is that the preparation of RSDF has started to influence the process of RPG formulation and the content of the various other regional strategy exercises. Although the degree of policy 'fit' varies between regions, and despite the fact that the timetable for the preparation of the various elements of regional strategy is not always helpful, it is clearly preferable to establish both the overall sustainable vision for a region – the job of RSDF – and the overall spatial strategy for a region – the job of RPG – at an early stage in the cycle of planning activities. In effect, this is what has happened in the East Midlands region through the preparation of a non-statutory Integrated Regional Strategy which can be used as a 'base' document for all other plans and strategies (East Midlands Regional Assembly, 2000). This example is discussed further in Chapter 11. In other regions a similar process is emerging, including the innovative approach adopted in Yorkshire and Humberside whereby a joint sustainable development appraisal has been conducted of draft RPG and the RDA strategy.

**Progress in Regional Environmental Planning**

The content and structure of the new environmental agenda for regional planning can now be defined and analysed in three ways:

- firstly, it is possible to examine and assess the 'internal' content and structure of the individual elements – RPG, RDA strategy, SDP, etc;
- secondly, the mechanisms for the 'horizontal' co-ordination and integration of the content of the various elements of strategy can be analysed within a region;
- thirdly, the various 'vertical' relationships that guide the content and structure of regional plans and strategies can be evaluated, such relationships reflect the characteristics and requirements of both 'higher' and 'lower' tiers of policy.

Figure 10.1 illustrates the connections between the various elements of regional planning and development. For the sake of clarity, only four individual elements of regional policy are included within the diagram – RSDF, RPG, RDA strategy and SPD. In reality, in addition to the plans noted above which incorporate some elements of environmental content, many other documents exist, including regional cultural strategies, regional plans for innovation and business support, regional tourism strategies, the regional co-ordination of Training and Enterprise Council programmes, regional and sub-regional technology plans, and so on.

The presence of the three components, or elements, of the new environmental regional agenda – 'internal' content and structure, 'horizontal' links and co-ordination within a region, and 'vertical' relationships between individual and co-ordinated plans in a region and the policies and priorities from 'above' or 'below' – provides both strengths and problems. For those who are engaged in the formulation of regional environmental planning and development policies, the strengths are:

- the presence of wide-ranging support for the acknowledgement and incorporation of (environmental) sustainable development objectives within regional plans, strategies and programmes;
- the existence of extensive scope for co-ordinated action and for the establishment of links between the various processes of planning and implementation;
- the presence of a range of activities that, when taken together, represents almost the full scope of environmental tasks and responsibilities that need to be planned and discharged at a regional level;
- the emergence of political structures that match and can provide support for the technical-professional tasks associated with regional planning and development;
- the possibility of going beyond the existing agenda, but this will depend on the conversion of the present opportunities into concrete arrangements and operational structures.

On the other hand, there are a number of difficulties that need to be addressed:

- the danger of actual and potential 'congestion' in terms of operational capacity at regional level due to 'initiative fatigue' and the lack of sufficiently trained and experienced professional regional planners and managers;
- the presence of a number of contested aspects of policy and territory as a result, in part, of the absence of elected regional authorities with adequate powers and resources;
- the continued use (in some cases) of unsuitable regions that are unlikely to prove to be satisfactory as operational territorial units;
- the influence and consequences of disputes within central government that may destabilise regional interworking and discourage co-ordination.

Having identified these strengths and difficulties, it is important to note that progress so far in achieving the goals of an environmentally-sustainable regional agenda has been significant. However, much of this progress has been made on the assumption that political power and real financial resources will follow. For example, whilst it is interesting to speculate on the future prospects for the new style of RPG, which will provide a spatial strategy for a region and also incorporate a regional transport plan, the acid test from a sustainable development perspective will be the extent to which the Regional Transport Strategy element of RPG accepts and expresses sustainable development priorities. This will be tested further through the overall spatial distribution and the modal allocation of transport-related public expenditure. The old maxim, 'to spend is to choose' still holds true, and a full assessment of the strength of the influence exerted by sustainable development on regional planning and development policy cannot be provided at present.

**Figure 10.1: Content and Structure of Regional Plans and Strategies**

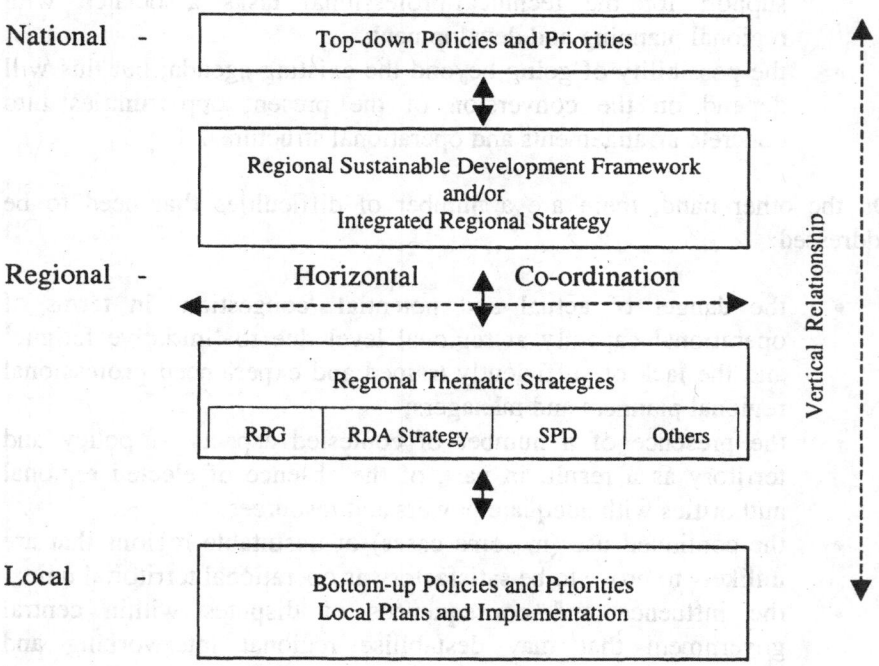

*Source:* Author

## Overall Models for Regional Environmental Planning

In all of the key areas of regional planning and development activity there are, as was seen above, a number of common features, such as the desire to reduce or make more efficient use of renewable and non-renewable resources, to control or eliminate pollution, and so on. In addition to these thematic elements of policy, there is also evidence of a growing concern with the development of a more robust and comprehensive guiding philosophy. Three elements of theory and practice that contribute to the provision of a basis for the formulation of a new operational philosophy have been considered in various regions:

- regional strategic environmental assessment – this is most clearly seen in the work now beginning in North West England that is built on the sustainable city-region demonstration project undertaken in Greater Manchester (Ravetz, 2000);
- regional industrial ecology – this is the subject of a number of small-scale 'experiments' in various regions, including the work on regional environmental technology, which has recently commenced in Yorkshire and Humberside;
- regional ecological modernisation – this is an approach to regional planning and development that has received little attention in the UK until recently, but it is now the subject of considerable interest in the East of Scotland (Jackson and Roberts, 1999) and is an approach that has been adopted by a number of Structural Funds regional programmes in 'mainland' European regions (Roberts, 1997).

Of these models, the first and third would appear to offer the greatest potential for use in regional planning and development, especially given that the fundamental elements of industrial ecology can be incorporated in an ecological modernisation approach (Mol, 1999).

There are three reasons for suggesting that a combined or hybrid model would appear to offer the greatest potential. The first reason relates to the inherent characteristics of the ecological modernisation approach. Chief among these characteristics is that the model recognises and accepts the importance of attempting to enhance environmental quality and bring about the more effective use of resources, whilst at the same time trying to incorporate 'the changes in modern society's institutions and practices that are relevant in safeguarding the sustenance base' (Mol, 1999, p. 170). This defining characteristic of ecological modernisation theory has a close resemblance to the essential features and characteristics of the 'balanced' regional planning and development model from the 1930s (Roberts, 1997). A second reason which justifies the advocacy in support of the hybrid model relates to the essential requirement that a model must be able to distinguish between different sectors and themes, on the one hand, and must be capable of providing a spatial perspective, on the other hand. It is now generally accepted that the specific economic, social and environmental conditions that are experienced in an individual place may necessitate the development of a specific 'blend' of policy (Gouldson and Roberts, 2000); such an approach matches the features of 'leading edge' practice in which sub-regional and local aspects of sustainable development are emphasised. The third reason supporting the use of the hybrid model is associated with the need to develop and advance the implementation of the model through a collaborative mode of interaction. As Ravetz has argued, a regional environmental planning and management framework would 'aim to be horizontally integrated between various public, private and quango sectors and activities; vertically integrated between regional, sub-regional and district-level bodies; and environmentally integrated between various physical media and pressures' (Ravetz, 2000, p. 199).

Although many of the ideas that are incorporated in this hybrid model are not new, the bringing together of environmental, social and economic and spatial processes does provide a basis for real co-ordination and integration. However, as Gibbs (1996) has argued, it is also important that a suitable mode of institutional expression and management is provided in order to ensure that a collaborative approach is adopted. Given the inherently destructive nature of many traditional forms of economic organisation, this suggests that the adoption of a collaborative approach is an essential requirement. As noted earlier in this chapter, this theoretical perspective has been confirmed by the empirical work undertaken by the UK Round Table on Sustainable Development (2000).

Finally in this section, it is important to be aware of two additional points:

- firstly, much of the new emphasis placed on the environment (together with a number of other aspects of sustainable development) has been added to the 'standard' agenda of regional planning and development – this issue is discussed in detail elsewhere (Green Alliance, 2000; Taylor, 2000; Roberts, 1999) and is not developed further here;
- secondly, many of these new aspects of spatial intervention policy at the regional level are seen by existing players as potential competitors for political and expenditure support. This has occurred despite the fact that a considerable body of evidence now exists which can be used to demonstrate the economic benefits that flow

from the adoption of an environmental approach to the preparation of a regional strategy.

## Moving Forward

The final section of this chapter attempts to take a step into the future and anticipates further elaboration of the regional agenda, especially in relation to the incorporation of environmental considerations and priorities. An assumption is made that the current English regional 'project' will continue and that the EU will take additional steps to strengthen the environmental component of its policy portfolio (European Parliament, 1998).

From January 2000, the operation of the Structural Funds has been guided by enhanced regulations that, among other matters, place greater emphasis on environmental performance than in the past. This, coupled with the introduction of Strategic Environmental Assessment and a significantly strengthened successor to the 5$^{th}$ Environmental Action Programme (Hamblin, 1999), should ensure that all regional plans and programmes associated with EU programmes will comply with a minimum level of environmental performance and, in the case of those programmes that wish to access EU funding, will have to achieve a much higher level of environmental performance than in the past. It should be noted that this raising of environmental standards will apply across the EU, irrespective of designated area status. Furthermore, it is also the case that an element of the resources made available to the countries of central and eastern Europe who are seeking accession to the EU are earmarked for improving environmental performance.

In the context of the EU requirement that all regional plans and programmes operate to a higher standard of environmental performance, the most important roles to be performed by central government are to:

- ensure that all areas of activity that relate to or influence regional planning and development are operating in accord with the relevant EU norms of environmental performance;
- encourage the greater co-ordination and integration of policy at central government level and through the guidance provided by the Government Offices – this suggests that the mandate of the GOs will have to be extended and that they will need to establish new relationships with the full range of regional actors;
- develop and apply performance indicators that reflect environmental priorities rather than the standard 'tonnage ideology' approach – this

requirement is now the subject of detailed discussion as part of the process of implementing RSDFs.

At regional level, the agenda over the short-term (the next two to three years) will chiefly be concerned with making the system work; especially with regard to the integration and co-ordination of the various aspects of spatial intervention policy and ensuring that the environmental dimension of sustainable development is incorporated into all of the various processes of planning and development. This, in ecological modernisation terminology, is an 'institutional learning' stage. In the medium-term (after the next election and up to 10 years hence), the priority will be on establishing detailed operational programmes that emphasise environmental priorities alongside the achievement of social justice and responsible economic progress. It is at this second stage that new theories, methods and operational approaches will be introduced in order to ensure that the best use is made of the available natural, human and financial resources, and to allow for the introduction of new mechanisms for the spatial integration of policy. During this second stage it is also anticipated that directly-elected regional authorities will assume responsibility for the regional planning and development agenda, and that they will work alongside other partners to ensure complete regional governance (Roberts, 2000). There are already signs that regional governance institutions place considerable emphasis on the achievement of sustainable development. Stage three (beyond 10 years) will see the adoption at EU level of an integrated sustainable development strategy for Europe as a whole which incorporates all aspects of policy and articulates the priorities for individual member states through a series of spatial management objectives and targets. Within the English regions this will help to complete the programme of change in political culture that started in the closing years of the twentieth century.

Although these possible pathways for the future evolution of the environmental component of regional planning and development may appear to be ambitious, they represent no more of a change than that which has been achieved during the past decade. Most importantly, these pathways help to embed the regional level of intervention within the institutional framework for the achievement of sustainable development. As was the case in Howard's analysis of the failure of conventional modes of spatial planning and management in the nineteenth century, the 'master key' to the achievement of sustainable development in the present era may prove to be the introduction of more effective regional planning, development and management.

# References

Ash, M. (1999), 'Regional structures – coherence or conflict?', *Journal of Planning and Environment Law*, Special Issue October, pp. 18–27.

Baker, M. (1998), 'Planning for the English regions', *Planning Practice and Research*, vol. 13, pp. 153–69.

Baker, M., Deas, I. and Wong, C. (1999), 'Obscure ritual or administrative luxury? Integrating regional planning and regional development', *Environment and Planning B*, vol. 26, pp. 763–82.

Breheny, M. (1991), 'The renaissance of strategic planning?', *Environment and Planning B*, vol. 18, pp. 233–50.

Carter, C. and Winterflood, S. (2000), 'Integrating environment into regional plans and strategies: the case of the West Midlands', in A. Gouldson, and P. Roberts (eds.), *Integrating Environment and Economy*, Routledge, London.

Chapman, M. (1998), *Effective Partnership Working: Good Practice Note 1*, Scottish Office, Edinburgh.

Cohen, M. (1993), 'Megacities and the Environment', *Finance and Development*, vol. 30, pp. 44–47.

Cronon, W. (1991), *Nature's Metropolis*, W.W. Norton, New York.

Department of the Environment, Transport and the Regions (1999), *RDA's Regional Strategies*, DETR, London.

Department of the Environment, Transport and the Regions (2000a), *Planning Policy Guidance Note 11: Regional Planning*, DETR, London.

Department of the Environment, Transport and the Regions (2000b), *Guidance on Preparing Regional Sustainable Development Frameworks*, DETR, London.

East Midlands Regional Assembly (2000), *Integrated Regional Strategy*, East Midlands Regional Assembly, Melton Mowbray.

European Parliament (1998), *Sustainable Development: A Key Principle for European Regional Development*, European Parliament, Brussels.

Friedmann, J. and Weaver, C. (1979), *Territory and Function*, Edward Arnold, London.

Geddes, P. (1915), *Cities in Evolution*, Williams and Norgate, London.

Gibbs, D. (1996), 'Integrating sustainable development and economic restructuring: a role for regulation theory?', *Geoforum*, vol. 27, pp. 1–10.

Gibbs, D. (1998), 'Regional Development Agencies and Sustainable Development', *Regional Studies*, vol. 32, pp. 365–8.

Glasson, J. (1995), 'Regional Planning and the Environment: Time for a SEA Change', *Urban Studies*, vol. 32, pp. 713–31.

Gouldson, A. and Roberts, P. (2000), 'Integrating Environment and Economy: The Evolution of Theory, Policy and Practice', in A. Gouldson and P. Roberts (eds.), *Integrating Environment and Economy*, Routledge, London.

Green Alliance (2000), *Sustainable Development and the English Regions*, Green Alliance, London.

Hall, P. (1974), *Urban and Regional Planning*, Penguin Books, Harmondsworth.

Hamblin, P. (1999), 'Environmental integration through strategic environmental assessment: prospects in Europe', *European Environment*, vol. 9, pp. 1–9.

Healey, P. (1997), *Collaborative Planning: Shaping Places in Fragmented Societies*, Macmillan, London.

Howard, E. (1998), *To-morrow: A Peaceful Path to Real Reform*, Swan Sonnenschein, London.
Hunt, Sir Joseph (1969), *The Intermediate Areas: Cmnd 3998*, HMSO, London.
Jackson, A. and Roberts, P. (1999), 'Ecological modernisation as a model for regional development', *Journal of Environmental Policy and Planning*, vol. 1, pp. 61–75.
Mol, A.P.J. (1999), 'Ecological modernisation and the ecological transition of Europe', *Journal of Environmental Policy and Planning*, vol. 1, pp. 167–181.
Murdoch, J. and Tewdwr-Jones, M. (1999), 'Planning and the English regions: conflict and convergence amongst the institutions of regional governance', *Environment and Planning C*, vol. 17, pp. 715–29.
Possiel, W., Saunier, R. and Meganck, R. (1995), 'In-situ conservation of biodiversity', in R. Saunier and R. Meganck (eds.), *Conservation of Biodiversity and the New Regional Planning*, Organisation of American States, Washington DC.
Ravetz, J. (2000), 'Strategic planning for the sustainable city – region', in A. Gouldson and P. Roberts (eds.), *Integrating Environment and Economy*, Routledge, London.
Roberts, P. (1994), 'Sustainable regional planning', *Regional Studies*, vol. 28, pp. 781–87.
Roberts, P. (1996), 'Regional Planning Guidance in England and Wales: back to the future', *Town Planning Review*, vol. 67, pp. 97–109.
Roberts, P. (1997), 'Sustainable development strategies for regional development in Europe: an ecological modernisation approach', *Regional Contact*, XI, pp. 92–104.
Roberts, P. (1999), 'Finding A way through the regional maze: the journey so far', Town and Country Planning Association Conference, Birmingham, January.
Roberts, P. (2000), *The New Territorial Governance: Planning, Developing and Managing the UK in an Era of Devolution*, Town and Country Planning Association, London.
Roberts, P. and Lloyd, G. (1999), 'Institutional aspects of regional planning, management and development: lessons from the English experience', *Environment and Planning B*, vol. 26, pp. 517–31.
Taylor, D. (2000), 'Integrating economic and environmental policy: a context for UK local and regional government', in A. Gouldson and P. Roberts (eds.), *Integrating Environment and Economy*, Routledge, London.
Tomaney, J. (1999), 'New Labour and the English Question', *The Political Quarterly*, vol. 96, pp. 75–82.
UK Round Table on Sustainable Development (2000), *Delivering Sustainable Development in the English Regions*, UK Round Table on Sustainable Development, London.

# PART IV:
# REGIONAL PLANNING –
# REGIONAL VARIATION

# PART IV

# REGIONAL PLANNING – REGIONAL VARIATION

# 11 Integrated Policy Development at the Regional Level – A Case Study of the East Midlands Integrated Regional Strategy

TONY AITCHISON[1]

## Introduction

This chapter addresses a number of key issues concerning the development of regional strategies and policies in the context provided by current regional institutional arrangements, roles and responsibilities and given the increasing interest in the regional policy agenda. It uses the East Midlands Integrated Regional Strategy (IRS) as a case study to demonstrate how it is possible for a 'voluntary' Regional Assembly to play a leading role in securing an integrated approach to regional policy development. Some benefits of an integrated approach for the development of regional planning guidance (RPG) are outlined.

## Background to the Preparation of an Integrated Regional Strategy

At its first meeting in December 1998 the East Midlands Regional Assembly[2] resolved to prepare an IRS. To understand why the Assembly regarded this as its key priority at its inaugural meeting we need to consider the background and context for this decision.

The new Labour Government elected in May 1997 pledged its commitment to create Regional Development Agencies (RDAs) to improve the competitiveness and prosperity of the English regions. The new agencies would be required to produce regional economic strategies. Many organisations in the regions, including the new Assemblies, were keen to influence the Agency and its strategic thinking.

As part of the European debate on 'Agenda 2000' concerning the future of European Structural Funds, European Commission officials were stressing that, in future, regions would need to demonstrate how European funding programmes helped to deliver the economic regeneration component of comprehensive regional strategies which addressed the full policy agenda. The previous approach of developing a series of initiatives and programmes of action in a regional policy vacuum would no longer be acceptable to the Commission.

As part of the preparations for the creation of the Regional Assembly, a number of partners commissioned consultants to carry out an analysis of all documents with a policy or programming content at regional or sub-regional level (ECOTEC, 1998). The consultants were also asked to obtain the opinions of key players on the proposal to prepare an IRS. Over 30 documents were analysed with some degree of strategic policy content. The consultants found that the policy coverage reflected the varying purposes for which the documents had been drafted. They included bidding documents for European funds, policy frameworks for allocating domestic resources, partnership-based sectoral strategies, single agency lobbying and promotional strategies, regional delivery or operational plans for the implementation of national programmes and business plans. No single strategy covered the full spectrum of economic, social and environmental issues. Social issues tended to be considered as a means of achieving economic objectives (skills development) and whilst social inclusion was a widely held objective, no document took a comprehensive approach towards this. Environmental issues tended to be treated as constraints. There were very few genuine partnership based regional strategies, even in the economic field, and those strategies identifying geographical priorities were not consistent. Policy documents and strategies often addressed different time scales and parts of the region.

The picture is one of confusion with *ad hoc* policy development being undertaken by a wide range of organisations, agencies and groupings, often with little knowledge or reference to other related policy work or strategies. This is not surprising, given the institutional context with no regional tier of government and after a long period during which government was predominantly concerned with individual programmes of action and the strengthening of central control mechanisms. It does, however, present a considerable challenge to those who can see the value of joining initiatives up at the regional level. Fortunately, the majority of regional partners fell into this category and supported the proposal to develop an integrated regional strategy. The consultant's findings presented the new Assembly with a robust analysis of disarray in the policy field it was about to join as a new player.

Various studies have listed the wide range of quangos and government agencies that take funding and programming decisions which have a significant impact upon the region, but with little or no regional involvement. Harding (2000, p. 22) writes that: 'The broad picture is that there is not so much a "missing middle" in English governance but rather a collection of different sorts of organisation, developed for different purposes at different times and by different levels of government, which forms a system of near Byzantine complexity. What is missing is any sense of collective purpose or coherence that clearly unites the various parts of the whole.'

For many years these organisations have operated comfortably within their own 'silos', often making decisions without reference to other organisations or government agencies operating in the same field. David Marquand and John Tomaney (2000, p. 7) contend that: 'As Labour noted in opposition, a tier of government already exists in England. The fact is that this tier is fragmented and poorly co-ordinated, organised along functional rather than territorial lines, has led to poor policy outcomes and a lack of joined up government in the regions.' Analyses like these helped to define the scope of the challenge which faced the East Midlands Regional Assembly.

The view of the new Government was that the development of regional economic strategies by the RDAs would be the primary focus of regional policy-making. For many years local government in the region had been very active in developing and reviewing a regional land use planning strategy which included strategic planning, transportation and economic development policies (East Midlands Regional Planning Forum, 1992). This was prepared to influence and inform RPG to be issued by the Secretary of State. At an early stage, the Government reaffirmed the importance it attached to regional planning bodies continuing with this role.

There was no requirement or expectation on the part of Government that other aspects of regional life, such as environmental or social issues, would be the subject of policy development at the regional level. This is a reflection of the centralisation trend and contrasts with the increasing recognition that at the national and local levels a holistic approach is needed, for example advice concerning the preparation of Community Strategies (DETR, 2000a). The European Commission had long recognised the significance of achieving social inclusion as a policy objective for regions and the importance of the wider social agenda including lifelong learning. Might there also be benefits in achieving integrated policy development at the regional level?

Although the Regional Assembly would not be regional government, it believed that it could perform a valuable function in bringing together the key regional policy and decision-makers, on a voluntary basis, to try and secure a more integrated and coherent approach to the 'new' regional policy agenda. The view of Regional Assembly partners was that this agenda must address environmental, social and spatial issues alongside the economic strategy to be prepared by the RDA. The Assembly therefore developed a diagram to illustrate its thinking about how these key areas of regional policy could be developed and delivered (Figure 11.1). The key feature of this early diagram is the emphasis placed upon *achieving integration* between the social, economic, environmental and spatial themes through policy development at the regional level.

Another early diagram (Figure 11.2) illustrated the complexity of the 'challenge' in trying to achieve a genuinely integrated regional strategy – if all the linkages with other relevant strategies and policies at different spatial levels are to be properly taken into account.

## The Rationale for and Approach to the Development of an Integrated Regional Strategy

In the absence of regional government and any responsibility for the development of an overarching regional strategy, the Regional Assembly set out to develop mechanisms to encourage an integrated approach to regional policy development on a *voluntary basis*. An overall vision for the region has been agreed by all regional partners including the East Midlands Development Agency (emda). Eighteen regional sustainable development (SD) objectives have also been agreed for the social, environmental, economic and spatial themes. The vision and SD objectives are included in Annex 1. Together they provide the overall framework for the development of regional policies. It is intended that all policy development at the regional level will be driven by these objectives and also appraised against them. Individual strategies, which are the responsibility of different agencies and organisations, form the *component parts* of the IRS. For example, the economic strategy developed by emda comprises the economic component of the IRS. The regional spatial development strategy developed by the East Midlands Regional Local Government Association (EMRLGA) as draft RPG comprises the spatial component of the IRS. The Regional Assembly does not therefore prepare its own 'competing strategies' but works to secure integration and compatibility between strategies developed by regional bodies and agencies. Although a wide range of bodies and agencies have environmental remits, no single body or organisation has responsibility for the development of a regional environment strategy. The Assembly has therefore taken the leading role in co-ordinating the development of a draft environmental strategy for the region, but with the active involvement of all the relevant agencies (Environment Agency, English Nature, Countryside Agency and so on) as well as the key environmental interests.

## Figure 11.1: Regional Strategy/Policy Framework

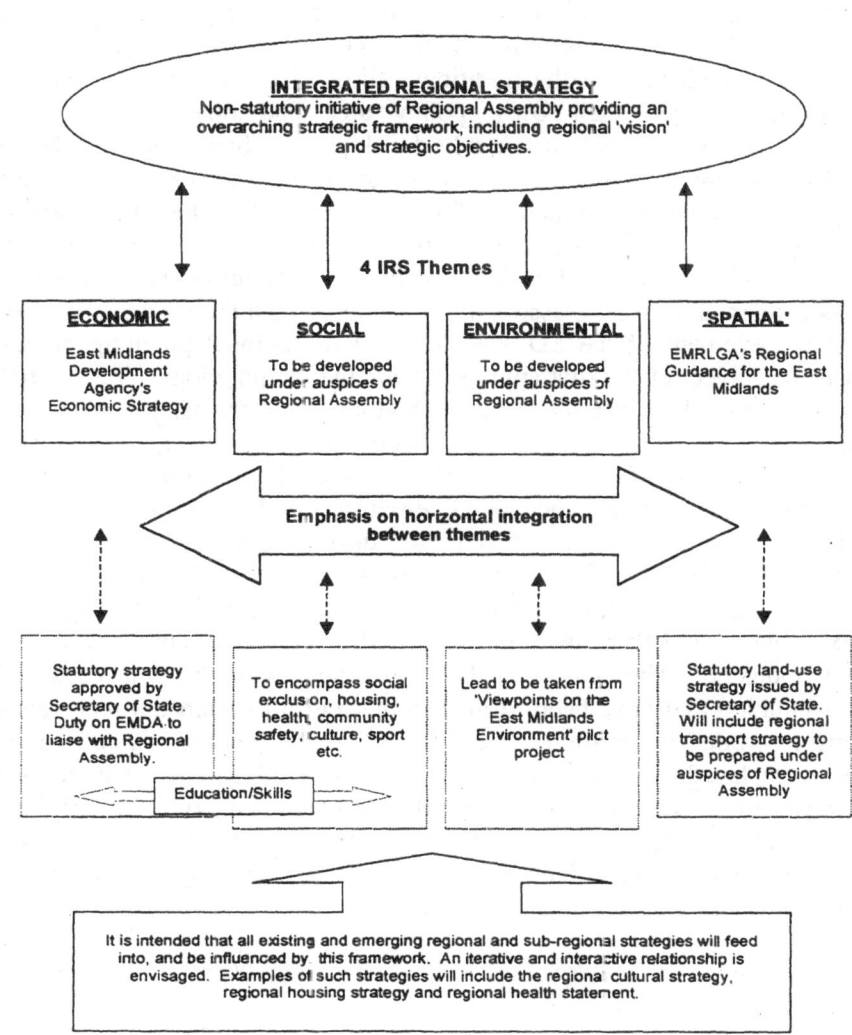

Source: East Midlands Regional Assembly, 2000c

This is a new approach to policy development at the regional level and the Assembly recognised the need to provide guidance to its regional partners. It therefore prepared a guidance note to assist policy and decision-makers which outlines the overall approach to the development of integrated regional policies and provides a number of aids for policy makers (East Midlands Regional Assembly, 2000a). The guidance note urges policy makers to look beyond the traditional silos, to consider the important linkages between their own policy area and other key policy areas and to use the regional sustainable development objectives as a checklist throughout policy development. A key aid to policy makers has been the development and dissemination of a consistent methodology to undertake a sustainability appraisal of regional policies and strategies (East Midlands Regional Assembly, 2000b). The methodology requires policy makers or independent assessors to assess the contribution that their strategy makes to the achievement of the SD objectives. Draft regional planning guidance was the subject of the first appraisal using the methodology. The Assembly persuaded emda to use the same methodology and the same consultants to appraise the economic strategy. Revisions and improvements were made to both strategies to address issues identified through the appraisal process.

The importance of the social, environmental, economic and spatial objectives in underpinning this policy development approach is crucial. The Assembly has acknowledged, however, that given the 'newness' of much of this regional policy work, the objectives should not be regarded as fixed but should be refined and reviewed as part of the process of policy development. The objectives have therefore been the subject of wide consultation and have been refined in the light of comments received and further work. For example, the three original economic objectives have been replaced by five more comprehensive objectives, which derive directly from work on the economic development strategy. Similarly, it has been acknowledged that the current spatial objectives will need to be reviewed when regional planning guidance is issued by the Secretary of State.

**Figure 11.2: The Challenge of Achieving Integrated Policy Development at the Regional Level**

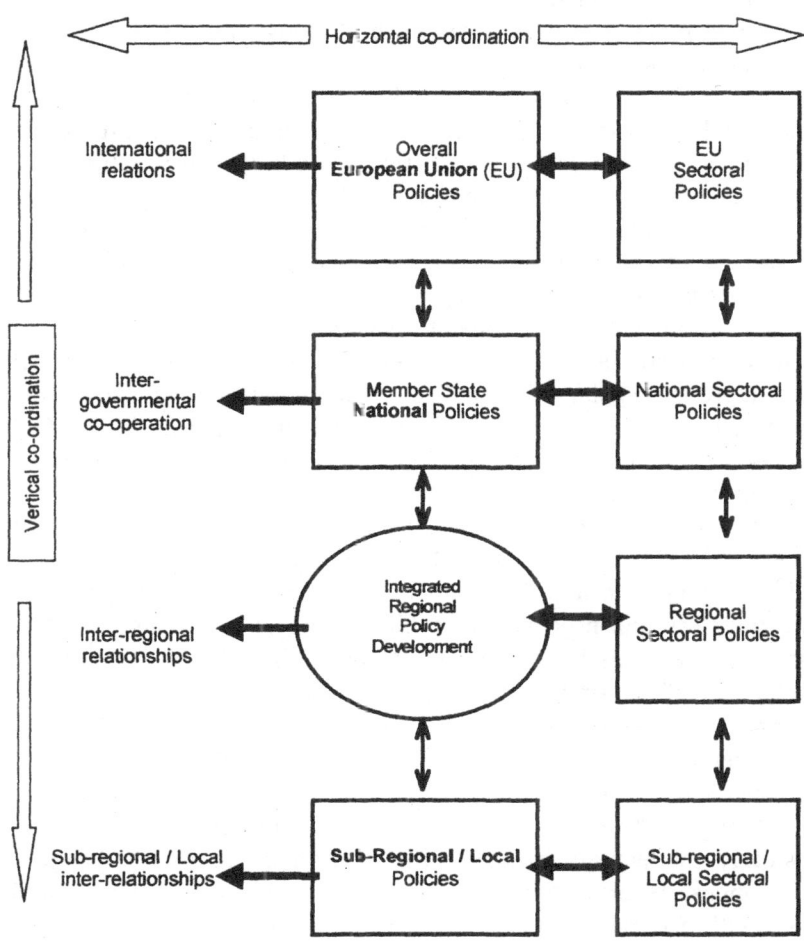

*Source:* East Midlands Regional Assembly, 2000c

## Collaborative Working on the Development of Regional Policies and Strategies

There is currently no requirement on any single regional body to prepare a regional strategy that addresses the full regional agenda in a holistic manner. Roles and responsibilities have evolved over time in an *ad hoc* manner. In some policy areas there are clear responsibilities with a single agency, whilst in others, various agencies, government departments and voluntary groupings are urged to work together, often based upon guidelines issued by central government. A few examples will illustrate the current diversity and complexity and difficulties which these institutional arrangements can present to those seeking to achieve a holistic approach.

RDAs have responsibility for developing regional economic strategies and associated delivery and action plans. The newly created Cultural Consortium is charged with preparing a regional cultural strategy. The overarching sustainable development framework is not the clear responsibility of a single body, but all relevant parties such as regional chambers, regional sustainable development round tables and the Government Offices (GOs) are encouraged to work together. The GO currently has the lead role for the development of the single programming documents for Objective 2 European structural funds. The RDAs have a secondary role, but the intention is that this should increase over time. The development of RPG is started by the Regional Planning Body, a voluntary organisation, is subject to independent scrutiny by a government appointed panel, and transfers mid-process to the GO before the Secretary of State finally issues guidance as his guidance. Responsibility for developing the regional chapters of the English Rural Development Plan lies jointly with the Ministry of Agriculture, Food and Fisheries (MAFF) and the GO.

Although central government created most of these arrangements, it had not seen a need to ensure that mechanisms are put in place to achieve integration between these interrelated policy and programming activities. This has clearly not been an imperative for government until quite recently. In many instances, the primary driver will be the need for compatibility and consistency with national policy or the way it is applied within the region. Guidance to those developing regional policies and programmes often urges them to take account of related work and issues. In reality, as far as Government is concerned, the most important test of any policies developed at the regional level, such as RPG, has been compatibility with national policy.

Devolution measures and the creation of RDAs, Regional Chambers, and organisations such as the Cultural Consortia have changed the regional dynamics. These organisations wish to identify the needs of the region and develop policies and programmes which are customised to regional circumstances. This new dynamic has created a climate within which new regional organisations can see the value of working to an agreed vision and set of objectives and of working collaboratively. In the East Midlands, the IRS has provided the mechanism and symbol of this desire for closer collaborative working amongst regional agencies and players.

This may seem to be only a theoretical ideal, but by taking as an example the key theme for this book, RPG, it is possible to demonstrate how it can also work in practice. Collaborative working on the development of RPG and the economic strategy is strongly recommended in government guidance. The EMRLGA, as regional planning body, and emda agreed that the economic strategy and RPG should be component parts of the IRS and be developed within the overall framework it provides. The time available for the development of the economic strategy between spring and autumn 1999 was very tight and coincided with the period when draft RPG was being prepared. Close and collaborative working on the two strategies included exchanging drafts of text and policies on an informal basis. Other means included joint commissioning of studies, regular liaison between emda board members and the senior members of the Planning Forum, close liaison between the Directors of Strategy of the two organisations and joint project groups on specific and contentious issues. It was agreed that the final drafts of both strategies should be the subject of sustainability appraisals to be undertaken using the same methodology and by the same consultant. It is worth repeating that the methodology used required that both strategies should be assessed against the IRS SD objectives.

Clearly the responsibility for approving draft RPG lies with the regional planning body, and for finalising the economic strategy with emda. However, given the desire to work collaboratively and to secure integration, it was agreed that the final drafts of both strategies would be considered at the same meeting of the Regional Assembly in September 1999. The Chairs of both organisations presented the draft strategies and sustainability appraisals, which were then the subject of close scrutiny within Assembly workshops. Members debated the strategies using a set of questions to probe the degree of compatibility and integration between them.

The 'Guidance Note' for policy-makers (East Midlands Regional Assembly, 2000a) includes a key principle that strategies should be developed with the full involvement and collaborative working between regional stakeholders and agencies. The Assembly's IRS policy forum, which steers the development of the IRS, has therefore been set up to include representatives from all key sectors as well as the GO and emda.

Task groups for transport, social inclusion, housing and the environment advise the policy forum. These have also been set up to achieve inclusive membership, for example, the housing task group, chaired by an Assembly representative for social landlords, includes representatives from housing associations, local government, the Housing Corporation, the health sector, private house builders, mortgage lenders, tenants representatives and so on. Similarly, the transport task group includes representatives from the rail industry, rail users, bus operators, cyclist groups, environment groups, business and local government. It is advised by the Highways Agency, Strategic Rail Authority, GO and local government transport specialists. To remain manageable, task groups have been kept relatively small. The Assembly has therefore set up larger reference groups which meet less frequently for wider consultation and to inform and test the task group's thinking with the full range of regional partners. In this way the Assembly seeks to be inclusive and to try and achieve regional ownership of the developing sustainable development and policy agenda.

In addition, regular high-level liaison meetings take place between the key regional agencies to facilitate discussion about policy and programming issues. Regular liaison meetings take place between Regional Assembly and emda, between emda and the Regional LGA, between the GO and Regional Assembly and so on.

**Key Stages in the Development of an Integrated Approach to Regional Policy Development**

It would not have been appropriate or feasible for the Assembly to agree the methodology for developing an integrated regional strategy at its inaugural meeting. The methodology has been developed through discussion, debate and practical experience and has been steered by the Assembly's policy forum. The overall approach and key stages are demonstrated in Figures 11.1 and 11.3.

The first and crucial step, which does not appear on any diagram, was to secure all key regional players' agreement that an integrated approach to policy development was required. The overall framework for policy development is provided by the social, economic, environmental and spatial themes with strong emphasis on the importance of integration between themes. The next task was to define the key regional issues that need to be addressed within each theme. This led relatively quickly to the development of a set of draft regional objectives, which were subject to wide regional consultation before an agreed set of objectives emerged.

**Figure 11.3: Key Elements and Stages in an Integrated Approach to the Development of Regional Policies and Strategies**

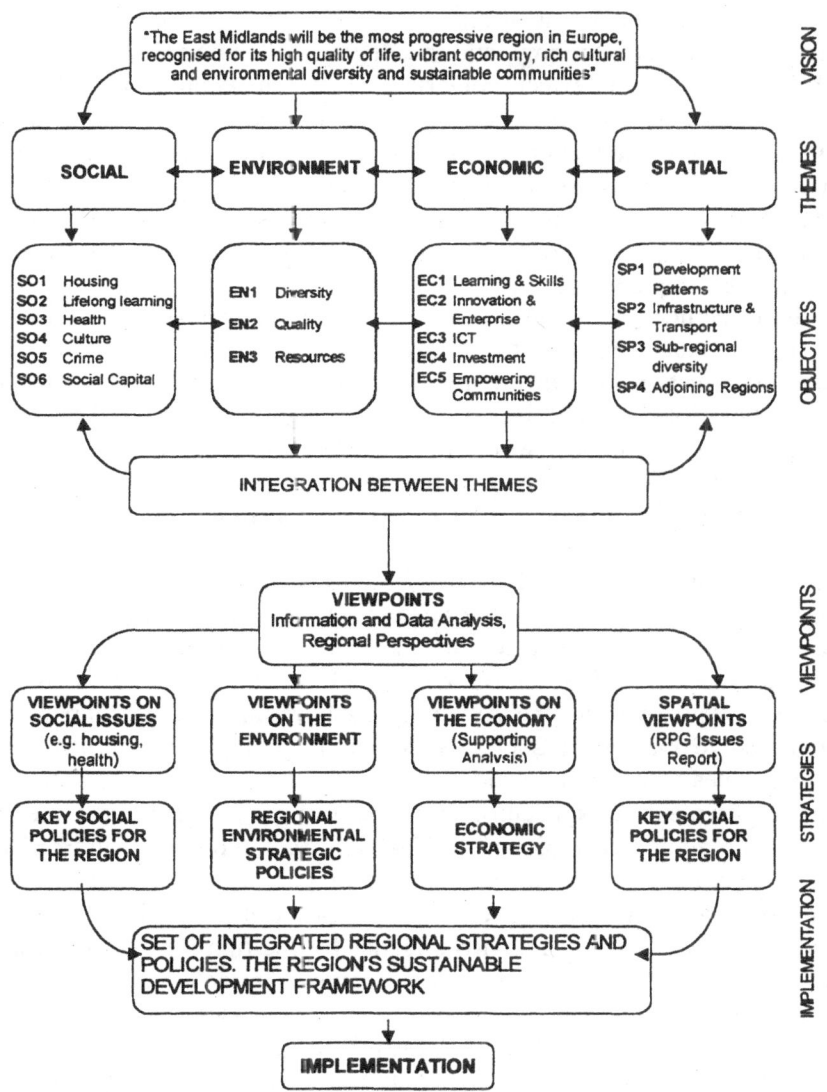

*Source:* East Midlands Regional Assembly, 2000c

The lack of robust analysis of the key issues, problems and challenges facing the region meant that a number of regional position statements were needed. These came to be known as the 'viewpoints' and the first viewpoints document concerned the regional environment. Emda commissioned an analysis of the region's economic position that comprised viewpoints on the economy, and the Regional Planning Body published an issues report that, in effect, comprised spatial viewpoints. The viewpoints on social exclusion document was published in December 2000 to complete the set. These viewpoints documents provide important information and analysis and a firm basis for the development of policies that can truly address the key issues and challenges facing the region. These policies, in turn, provide the basis for the action, delivery and business plans needed to secure their implementation.

In December 2000, just two years after its formation, the Regional Assembly has been able to approve the first edition of its IRS as the Regional Sustainable Development Framework (East Midlands Regional Assembly, 2000c). The IRS sets out the regional vision and sustainable development objectives, which provide the overall framework for policy development. For the first time the Assembly has brought together in a single document the key issues, current position on regional policy development and challenges in achieving genuinely integrated policy development and action. This first edition of the IRS acknowledges that policy development is not at the same stage for each of the key themes. The economic strategy is in place and regional planning guidance is nearing completion. The first consultation draft of the regional environment strategy, developed by the Assembly's environment task group was also approved in December 2000. Significantly, policy development on key social issues is the least developed and this can be related to the absence of any requirement or expectation on the part of government that these issues should be addressed regionally. The IRS does, however, summarise the latest position on the development of regional perspectives on social inclusion, housing, lifelong learning and skills, health, culture, and crime and community safety. The IRS includes for the first time a set of regional indicators and targets to measure progress in achieving the regional sustainable development objectives. It also stresses the significance of action at the sub-regional and local levels in achieving the implementation of many regional policies and programmes. This is considered further in the next section.

## The Benefits of an Integrated Approach to Policy and Strategy Development

The overall approach outlined above has achieved the full support of a wide range of regional players – but why? This section outlines some of the perceived benefits that an integrated approach to regional strategy and policy development brings. The agreement of a regional vision and set of regional objectives provides a unifying framework to drive all policy development and a mechanism for assessing their contribution to the achievement of regional aspirations.

Treating the economic and spatial strategies (namely RPG) as component parts of the IRS has led to collaborative working, as described above, which by external assessment has resulted in strategies with a high level of compatibility. The independent panel commented favourably upon this compatibility in its final report after the public examination into RPG. The Government urged each region to prepare a sustainable development framework to complement and help to apply the national sustainable development strategy. The Assembly was satisfied that its IRS already met the requirements of a sustainable development framework. When Sir Jonathon Porritt, Chair of the Sustainable Development Commission, and the Prime Minister's chief adviser on sustainable development matters addressed a conference on sustainable development in the region he said: 'I was intrigued, and I must say, very envious to hear about the way in which the East Midlands Regional Assembly has been able to use its *Integrated Regional Strategy* process to substitute for the development of a stand-alone *sustainable development framework* for the region. Unfortunately, we have no such luxury in the South West. After the regional planning process was nearly finished, and after the South West RDA's economic strategy was finished, we then started work on the Sustainable Development Framework. Of course you can understand that this is really totally the wrong way round. In a better regulated world, one would have as the over-arching document, the Vision, the framework for policy – the Integrated Regional Strategy as you call it in the East Midlands. From that over-arching strategy would flow those strategies specific to different regional entities, whether it was emda, the Government Office or the Assembly.'

The IRS framework provides a coherent context and rationale for taking forward various initiatives at the regional level. For example, regional partners have developed Social Viewpoints to identify the incidence and nature of social exclusion across the region. When set within the IRS framework this can be used as a key document to influence policy development and decision-making across the region. It can be promoted as an important component of a coherent process rather than an isolated one-off report. The way this document has been developed and presented has been influenced by its positioning within the IRS.

The IRS stresses the importance of linkages between regional, sub-regional and local policy development and decision-making. The feedback from those developing community strategies has been positive and in particular the IRS has been supported and welcomed as a valuable regional context and checklist for community strategy development.

The guiding principles that underpin the IRS and challenge silo thinking have resulted in different and refreshing approaches to policy development and delivery. For example, it has been more clearly appreciated that the health service cannot improve the region's health alone. All organisations can contribute to improving health through their own activities, programmes and work practices. Health representatives are engaged in the policy development being undertaken within Assembly Task Groups and are helping to highlight the key contributions which can be made through transport policy, housing measures and the working environment. Government policy guidance urges policy and decision-makers to be aware of these key linkages, but how can this be facilitated without the support and involvement of other relevant professionals? The IRS approach identifies the key issues beyond the normal silos which need to be taken into account. It also enables experts from the related fields to work collaboratively to advise upon the fields of others.

When the IRS has been the subject of consultation, responses have been predominantly constructive and have produced refinements that have helped to shape and influence its development. Clearly individual sectors and groups have often urged greater emphasis upon their particular concerns but this has been matched by support for the overall IRS approach.

Whilst the inclusion of the spatial theme distinguishes the IRS from the national sustainable development strategy, this has a number of benefits. Policy and decision makers are required to assess the spatial implications of their policies and programmes. This can be seen as a positive contribution to a greater appreciation of the importance of spatial considerations, as envisaged by the European Spatial Development Perspective. The land use planning system is a blunt instrument in influencing many public investment decisions, which will not even constitute 'development'. For example, closure, rationalisation and reorganisation of health facilities can have fundamental implications for spatial patterns, accessibility, social exclusion and so on. If the policy and decision maker is not required to consider these implications they will be ignored or secondary. However, if every agency and partner, such as those in the health sector or those receiving public funding via emda, have to consider the spatial implications of their actions through testing projects and programmes against the region's SD objectives, then this can only help to raise the awareness and significance of spatial considerations.

## Challenges in Achieving Genuinely Integrated Policy Development and Action Within the Region

As outlined above, responsibility for policy development and programme delivery in the region rests with a number of different government departments, agencies, quangos and bodies. The primary accountability of the GO and quangos will be to central government and individual Secretaries of State. Responsibility for national programme delivery and control of budgets currently lies with government departments, the GO and organisations such as emda and the new Learning and Skills Councils. Central government has recognised through the recently published Performance and Innovation Unit report, *'Reaching Out'* (Cabinet Office, 2000a) and the Regional Co-ordination Unit's *'Reaching Out – Action plan'* (Cabinet Office, 2000b) that it needs to achieve greater 'joining up' in the delivery of national policies and programmes within the region. The creation of a special unit within the Cabinet Office, the Regional Co-ordination Unit, to improve the integration of government delivery tends to confirm that civil servants and government see the primary means of achieving better joining up within the regions as being co-ordination at the centre. This contrasts with the East Midlands IRS approach, outlined above, which seeks to achieve better joining up *within* the region. Notwithstanding centralist tendencies within Government, the approach being developed within the East Midlands has been warmly welcomed by Hilary Armstrong, Minister for Local Government and the Regions. When addressing the Regional Assembly in April 2000, she commented that: 'Your Integrated Regional Strategy is about making sure that all the pieces of the East Midlands jigsaw fit together and develop in a coherent way, the wider regional agenda. This is an ambitious undertaking. But the framework you have developed to take forward the IRS seems to present the opportunity for all regional players, to work together towards a common vision and to ensure their particular piece of the jigsaw fits in with the overall picture. I am pleased that both the Regional Economic Strategy and the Regional Planning Guidance are intended to sit comfortably in this model. The Structure of the IRS – covering as it does the economic, social, environmental and spatial themes, makes it an ideal vehicle for the Sustainable Development Framework, which we have asked each region to produce. Again, the East Midlands approach is unique and I will be interested to hear how it develops. But clearly sustainable development needs to be at the core of all strategic thinking.'

The management team within the GO has been supportive of the IRS approach and has encouraged active participation particularly by senior members of staff in the collaborative working arrangements, both within the Assembly's Task Groups and on other initiatives within the IRS Framework. It has proved to be more challenging to try to ensure that other relevant staff are aware of the latest position on the IRS and the important contribution they can make through their own work to its development and delivery. To assist this process the Assembly organised a teach-in for all GO section heads to outline and discuss the IRS approach. The significance of loyalties and responsibilities to separate Government Ministers should not be underestimated, particularly in reinforcing a silo approach within which the primary measures of achievement are against 'departmental' targets.

Support for the IRS approach by the emda Board and senior management has been very strong. This has been demonstrated through their willingness to committ staff resources and funding to the joint development of policies and studies and active involvement on the Assembly's Task Groups. A sterner test of this approach is whether the individual programmes of action developed by emda, such as Single Regeneration Budget and rural action programmes are applying the SD approach down to project level. Are opportunities to foster social inclusion and to enhance the environment being increased through public funding support disbursed by emda to individual projects? Recognising this particular challenge, the Assembly has developed a sustainability checklist. This was developed from the regional sustainable development objectives and poses a series of questions to enable project developers and assessors to consider the contribution which individual projects can make to the achievement of the SD objectives (East Midlands Regional Assembly, 2000d).

In common with a number of other RDAs, emda are looking to rely upon sub-regional partnerships to deliver its economic programmes and initiatives. It is therefore imperative that the holistic approach represented by the IRS is also applied at the sub-regional and local levels. The Assembly has consistently stated that IRS is not a top-down strategy and has recognised the need for local community strategies and sub-regional thinking to influence the development of the regional policy framework.

In contrast to the economic strategy, which provides the context for the delivery of a wide range of economic regeneration programmes, the draft regional environment strategy, developed by the Assembly, is non-statutory and has been developed on a 'voluntary' basis. There is no equivalent agency to emda, such as the Environment Agency, with primary responsibility for preparing a regional environment strategy. Nor does a single agency have overall control of funding streams to influence all players to take positive action to improve the region's environment. This presents the Assembly and its partners with a challenge. How can an overarching regional environment strategy influence action across the region? This challenge has yet to be addressed, but one approach would be to develop a series of action plans for the individual strategy components (air, water, and so on) that set out the key contributions which different agencies, partners and organisations can make. Partners and agencies would need to sign up to their contributions to bring the strategy to fruition. If the key quangos continue with their active involvement and are willing to take a leading role in their particular field for the co-ordination and delivery of an overall action plan this would directly link such action plans to their funding regimes and priorities.

Whilst the IRS approach seeks to achieve synergy between regional policies and statements and to identify the positive linkages and benefits of integration, there is also a need to consider how tensions or conflicts are managed. The ideal is clearly to avoid conflicts by joint and collaborative working, and by the early identification of likely tensions. The IRS identifies the real challenges in achieving genuinely integrated action for each policy field. For example, under the spatial theme, the key challenge is 'how to accommodate growth in sustainable locations and to maximise the use of previously developed land, without leading to over-concentration in the three cities and northern coalfields sub areas'. A challenge in developing a regional housing strategy is 'the need to link with RPG to achieve a consistent approach to assessing overall housing needs and specifically for social housing'. A challenge which relates economic and environmental aspirations is 'to ensure that in seeking to push into the top twenty economic regions in Europe, the East Midlands must also ensure that our environmental objectives are achieved'. By highlighting these key challenges in a clear and open manner the IRS enables policy and decision makers to address them through their own work.

A primary role of sustainable development appraisals, when undertaken properly throughout the policy development process, is to identify and help to address such tensions and challenges. Similarly at the project level the use of the sustainability checklist should ensure that all pertinent factors are taken into account and given due weight throughout the decision-making process.

If we are to achieve genuinely integrated decision-making with the region, all these mechanisms need to be deployed to try to avoid conflicts and tensions. Where differences of view do arise, the IRS approach means that close and collaborative working should still be employed. For example, the undoubted locational advantages for future business development of the area around the East Midlands Airport resulted in divided views about the best course of action. The response was to commission independent consultants to analyse the pros and cons of developing in the area. A partnership-based working group was then assembled with all the key regional players including emda, the Department of Trade and Industry (DTI) part of the GO, the regional planning body, together with key local Councils, the business community, Highways Agency and so on to see whether a consensus could be reached on the way forward. The process of bringing such a group together and discussing the key issues in an open and transparent manner resulted in a better understanding of the issues, different viewpoints and potential ways of resolving these. The Group were able to agree a joint statement which was submitted to the RPG Public Examination. The next section considers other benefits of the IRS approach in developing a regional spatial strategy.

## The IRS Approach and RPG

The spatial strategy, RPG, is one of the component strategies and therefore an integral part of the IRS. Some have questioned whether there should be a spatial theme to the IRS. They argue that as any spatial strategy inevitably addresses social, economic and environmental matters, it should be regarded as an integrating strategy in its own right. Others, however, have argued that an economic development and regeneration strategy inevitably addresses social issues, such as skills development and environmental concerns, such as the removal of dereliction. They therefore regard the economic strategy as the core regional strategy. It is an unhelpful debate to try and determine pre-eminence between individual strategies. Such arguments tend to reinforce the value and need for an integrated approach, which requires all policy and decision makers to look at all the key linkages between topics and issues as an integral part of policy development. This section highlights some of the ways in which the IRS approach has influenced and benefited the development of the Regional Spatial Strategy (RPG).

In some regions policy development on RPG, the economic strategy, regional sustainable development frameworks and social policy issues is undertaken by different groups working to different bodies without any clear form of co-ordination. Jonathon Porritt's comments suggest that this way of working is not proving to be a coherent approach for the South West. In the East Midlands, the RLGA has a small policy team which supports and helps to co-ordinate policy development undertaken by the

Assembly's task groups and policy forum on many aspects of the IRS. The team also has responsibility for the co-ordination of the development of RPG. It liaises closely with emda on the economic strategy and provides advice to the sustainable development round table on the overall sustainability agenda. The involvement of the same individuals in advising on policy co-ordination in all these fields has helped to facilitate collaborative working and improved the coherence of policy development. This can be illustrated by focusing on policy development in two of the key themes of the IRS – the economic strategy and draft RPG.

The Regional Planning Body (RPB) and emda identified a number of key regional issues and projects where joint working and commissioning of studies would be beneficial. For example, studies have been undertaken by consultants to investigate ways of maximising urban capacity for house building, to assess the most suitable locations to attract major single user inward investors, and to identify the demand and supply of strategic high quality economic development sites. Generally the RPB has contributed resources 'in kind' through staff time and the agency has provided funding. They have been steered by joint steering groups with involvement of other partners as appropriate. It is fair to mention that during the development of draft RPG officials from the GO also worked collaboratively within many of the officer and advisory groups and indeed attended meetings of the Planning Forum. As outlined above, the contentious issue of economic development in the area surrounding the East Midlands Airport was also the subject of a joint study with a range of other partners.

Joint working and agreement upon the action needed as a result of these studies has laid solid foundations for collaborative working on the development of the respective strategies. Informal drafts of policies and proposals were exchanged between policy staff and high level meetings were held between senior Members of the Planning Forum and emda board members to discuss a range of key policy issues.

As a result, when the Public Examination into draft RPG took place in June 2000, the EMRLGA and emda were able to submit a number of joint background papers setting out their agreed analysis and conclusions on a wide range of issues, for the benefit of the examination. Furthermore, on a number of the key issues identified by the Panel, the two organisations were also able to submit joint statements, which may have surprised and disappointed a number of participants, such as national lobbying bodies accustomed to making capital out of conflicts and inconsistencies in regional documents!

It has been mentioned that both strategies were subjected to independent sustainability appraisals using the same methodology and consultant, who found a high level of compatibility between the two strategies. This conclusion was also endorsed in the final report of the Independent Panel when they reported following the public examination into draft RPG.

The way in which the two strategies dealt with transport infrastructure investment can be taken as an example to illustrate the benefits of joint and collaborative working within the IRS Framework. In a number of regions RDAs set out their own priorities for transport infrastructure investment as part of their draft economic strategies and this resulted in tensions between the RPB and RDA. After close liaison and joint agreement, the first draft of emda's economic strategy highlighted the importance of transport infrastructure investment but noted that the Regional Assembly's Transport Task Group and RPB were working with emda's active involvement on a set of regional priorities, which the final economic strategy would take on board. Agreement upon the most appropriate means of dealing with key issues within the IRS Framework therefore avoided the kind of conflict and arguments over 'competing strategies' that has occurred elsewhere.

Although the authors of PPG 11 may not have been aware of the detailed examples outlined above, they nevertheless wrote in the section dealing with the RPG/RDA relationship that: 'The East Midlands provides an example of best practice where the regional economic strategy and RPG are two components of an integrated strategy co-ordinated by the Regional Chamber' (DETR, 2000b). It is hoped that this implicit acknowledgement of the benefits of regional ownership might cause civil servants and Ministers to think carefully before rewriting RPG for a region.

**The Regional Policy Agenda**

There is keen interest in the future development of the regional policy agenda in the East Midlands. The new regional dynamics and policy agenda have provided the catalyst for the development and shaping of regional networks. For example, the thematic approach of the IRS has influenced the organisation and development of the voluntary and environmental sectors to enable them to feed into Assembly Task Groups to maximise their influence. The strong desire of regional players to push the regional agenda further may cause disquiet in parts of Whitehall, where the regional level tends to be regarded as a mechanism for administering government funding and applying national policies without a regional strategic policy input. This can be exemplified in the housing field, where the Assembly's Housing Task Group has agreed that it wishes to prepare a Regional Housing Strategy. The central government view seems to be that funding for housing should be administered through the Housing Corporation and Government Office but that prioritisation of projects should be based on national criteria. In the East Midlands, the Assembly's

Housing Task Group (which includes representatives from the Housing Corporation and GO) has adapted the form of the Housing Statement from that expected in Whitehall to enable it to become a regional 'viewpoints on housing' document with the overall IRS framework. The Task Group is now embarking upon the next stage to develop a Regional Housing Strategy, as a component part of the IRS. The reaction of central government will be interesting.

The guidelines issued by DETR for the preparation of Regional Sustainable Development Frameworks urge each region to include targets for renewable energy resources. How can these be developed so that they are supported by sound analysis rather than being plucked from the air to meet the arbitrary deadline of the end of 2000 for SD Frameworks? The Assembly has agreed that the most appropriate mechanism to undertake this properly is through the development of a comprehensive Regional Energy Strategy. Work has therefore commenced on this and the targets included in the IRS are described as interim pending the development of this strategy.

The Regional Cultural Consortium has been asked to prepare a Regional Cultural Strategy. It has been agreed that this should be regarded as one of the component strategies of the IRS and the Chair of the Consortium has joined the Assembly's Policy Forum to facilitate an integrated approach to its development. The latest thinking on the development of the Cultural Strategy was presented to a meeting of the full Assembly and was the subject of group discussions involving Assembly members and advisers. The draft set out the key linkages with other IRS topics to ensure these are fully taken into account.

Section two referred to the European Commission's view that bids for European regeneration programmes and single programming documents (SPDs) should demonstrate that they are key delivery mechanisms of comprehensive regional strategies and not isolated initiatives. This has to be the ideal. Consultants employed to develop the SPD for Objective 2 funding for the East Midlands had to include a position statement on the overall regional strategic framework for the programme. They have advised that, having worked in a number of English regions, the East Midlands IRS provides the best ready-made strategic policy statement they had seen.

Regional strategies should not, however, be prepared automatically for every topic or policy area, but only where it has been agreed that policy development adds value at the regional level. A key test is the extent to which regional strategies or policies influence action at the regional, sub-regional and local levels. This can be in a number of ways, ranging from the setting of the regional context or checklist for policy development at other levels, through to more prescriptive or binding policies, for example RPG. Influence can also be achieved by clarifying how different players can make a contribution, for example through the development of action plans for the regional environmental strategy. When strategies are linked to the likelihood of receiving financial support, as in the case of the economic strategy and associated action and delivery plans, influence can be considerable.

In some policy areas there have been interesting discussions about the added value of a regional statement, strategy or set of policies. The lifelong learning agenda provides a good example. The debate has been influenced by the reality that the key levers of financial control and distribution, policy and curricular development and inspection regimes are predominantly at the national and local levels. The Assembly has recently agreed to set up a Learning and Skills Task Group to assess the scope for adding value at the regional level in this vital realm.

**Early Conclusions**

This chapter has outlined the approach adopted within the East Midlands over the past two to three years in response to the new regional agenda. Unlike some regions, the East Midlands has not pressed for directly elected regional government. The Regional Assembly acknowledges that work needs to be done in developing regional awareness and enthusiasm for regional issues. It has therefore consciously dedicated its energies to achieving a more integrated approach to regional policy development. Significant progress has been made in a very short period. The early indications are that synergy can be achieved with a clear policy framework, which has achieved regional ownership and influence when combined with a will to work collaboratively. This can be the case even in policy areas where a regional dimension is not expected or required (or perhaps even welcomed) in some parts of government. It is beginning to provide the region with a coherent and effective means of addressing key issues at regional level. It is not so much a case of 'watch this space' but 'watch this region'.

**Notes**

[1] Tony Aitchison is Assistant Director (Strategy), East Midlands Regional Local Government Association. Views expressed here are those of the author and do not necessarily reflect the policies of the East Midlands Regional Local Government Association.

[2] The East Midlands Regional Assembly is the partnership based body (described as Regional Chamber in the Act) recognised by the Secretary of State as the democratic body with a relationship with the East Midlands Development Agency (emda). About 70% of its members are nominated by the region's local authorities and the 30% represent the social partners including business, trade unions, voluntary and other sectors.

## References

Cabinet Office (2000a), *Reaching Out – The Role of Central Government at Regional and Local Level*, Performance and Innovation Unit, February 2000, London.

Cabinet Office (2000b), *Reaching Out – Action Plan*, Regional Co-ordination Unit, October 2000, London.

DETR (2000a), *Preparing Community Strategies – Government Guidance to Local Authorities*, DETR, December 2000, London.

DETR (2000b), *Planning Policy Guidance Note 11, Regional Planning*, DETR, October 2000, London.

East Midlands Regional Assembly (2000a), *Guidance note for the development of regional policies and strategies within the Integrated Regional Strategy*, East Midlands Regional Assembly, available at [www.eastmidlandsassembly.org.uk].

East Midlands Regional Assembly (2000b), *Step-by-step guide to sustainability appraisal*, East Midlands Regional Assembly, available at [www.eastmidlandsassembly.org.uk].

East Midlands Regional Assembly (2000c), *The English East Midlands Integrated Regional Strategy – our sustainable development framework*, The East Midlands Regional Assembly, available at [eastmidlandsassembly.org.uk].

East Midlands Regional Assembly (2000d), *The East Midlands Regional Sustainability Checklist*, East Midlands Regional Assembly, October 2000, available at [www.eastmidlandsassembly.org.uk].

East Midlands Regional Planning Forum (1992), *Regional Strategy for the East Midlands*, East Midlands Regional Planning Forum.

ECOTEC (1998), *Research Study – Integrated Regional Strategy for the East Midlands*, ECOTEC Research and Consulting.

Harding, A. (2000), *Is there a 'missing middle' in English governance?*, Report to the New Local Government Network presented to the Devolution in England conference July 2000.

Marquand, D. and Tomaney, J., (2000), *Democratising England*, report to the Regional Policy Forum, November 2000.

## Annex 1: Vision and Regional Sustainable Development Objectives

### Vision for a Sustainable East Midlands

The East Midlands will be the most progressive region in Europe, recognised for its high quality of life achieved through a vibrant economy, rich cultural and environmental diversity and sustainable communities. We will make progress through:

- **Enterprising and innovative business** that can compete in the global market place, driven by and rewarding the knowledge and talents of their people.
- **Communities that empower people**, are safe and healthy, combat discrimination and disadvantage and provide hope and opportunities for all.
- **Conserving and enhancing** the diverse and attractive **natural and built environment** and cultural heritage of the Region and ensuring prudent management of resources now and for future generations.
- **Sustainable patterns of development** which enable social, environmental and economic progress.

### Sustainable Development Objectives

| Social Objectives | |
|---|---|
| SO1 | To ensure that the housing stock meets the HOUSING needs of all parts of the community. |
| SO2 | To ensure that the delivery of a wide range of LIFE-LONG LEARNING OPPORTUNITIES is provided for all parts of the community. |
| SO3 | To promote, support and sustain HEALTHY COMMUNITIES and lifestyles. |
| SO4 | To maximise the contribution of ARTS, CULTURE, HERITAGE, MEDIA AND SPORT to the quality of life of the East Midlands |
| SO5 | To ensure commitment and co-ordinated action to SECURE COMMUNITY SAFETY and reduce crime. |

| | |
|---|---|
| SO6 | To support the development and growth of SOCIAL CAPITAL[1] across the communities of the Region. |

| Environmental Objectives | |
|---|---|
| EN1 | To PROTECT, improve and manage THE RICH DIVERSITY of the natural and built environmental and archaeological assets of the Region. |
| EN2 | To ENHANCE and conserve the ENVIRONMENTAL QUALITY of the Region including high standards of design and maximise the re-use of previously used land and buildings. |
| EN3 | To MANAGE the NATURAL RESOURCES of the Region including water, air quality and minerals IN A PRUDENT MANNER and to seek to minimise waste and to encourage re-use and recycling of waste materials. |

| Economic Objectives | |
|---|---|
| EC1 | To bring about EXCELLENCE IN our approach to LEARNING AND SKILLS – giving the Region a competitive edge in how we acquire and exploit knowledge, by creating a 'learning region' – with individuals and employers who value learning and a learning industry that is proactive and creative – leading, in time, to a workforce that is among the most adaptable, motivated and highly skilled in Europe. |
| EC2 | To develop a strong CULTURE OF ENTERPRISE AND INNOVATION, putting the Region at the leading edge in Europe in our exploitation of research, recognised for our spirit of innovation – and creating a climate within which entrepreneurs and world-class businesses can prosper. |
| EC3 | To use the global INFORMATION AND COMMUNICATIONS TECHNOLOGY revolution to create the capability for everyone in the Region – individuals and businesses – to use information and knowledge to maximum benefit. |

---

[1] The Commission on Social Justice defined Social Capital as consisting of 'the institutions and relationships of a thriving civil society – from networks of neighbours to extended families, community groups to religious organisations, local business to public services, youth clubs to parent–teacher associations, playgroups to police on the beat. Where you live, who else lives there, and how they live their lives – co-operatively or selfishly, responsibly or destructively can be as important as personal resources in determining life chances.'

| | |
|---|---|
| EC4 | To create a CLIMATE FOR INVESTMENT in which success breeds success – providing the right conditions in the Region for a modern industrial structure on a combination of indigenous growth and inward investment. |
| EC5 | To EMPOWER COMMUNITIES to create solutions that meet their needs – ensuring that everyone in the Region has the opportunity to benefit from, and contribute to, the Region's enhanced economic competitiveness, thereby supporting a socially inclusive region. |

| Spatial Objectives | |
|---|---|
| SP1 | To ensure that decisions about the distribution and location of activity are consistent with SUSTAINABLE DEVELOPMENT PRINCIPLES. |
| SP2 | To ENHANCE the Region's INFRASTRUCTURE, including maximising transport choice and exploiting opportunities offered by information technology. |
| SP3 | To recognise and RESPECT the DISTINCTIVE CHARACTERISTICS OF different PARTS OF THE REGION and the need for regional policies and actions to take account of these. |
| SP4 | To have full regard to the importance of LINKAGES between different parts of the Region and WITH ADJACENT REGIONS. |

# 12 Regional Planning in the West Midlands

JOHN DEEGAN

**The West Midlands Region: An Overview**

*Land and Population*

The West Midlands region (Figure 12.1) lies in the centre of England and has a population of 5.3 million. It covers an area of 13,000 square kilometres including a metropolitan core centred around the Birmingham conurbation and the separate City of Coventry, surrounded by a green belt. Beyond the green belt there is a substantial hinterland of smaller towns and rural areas and in the north of the region is a smaller conurbation centred on the City of Stoke-on-Trent.

The metropolitan core occupies around 9% of the land area (1,200 square kilometres), but houses nearly 50% of its population (2.6 million). The regional population has grown by just over 1% since 1991 and is still growing but, as in many other regions, the current projections assume continuing migration away from the urban core.

*The Regional Economy*

The West Midlands regional Gross Domestic Product is about 8.5% of total United Kingdom (UK) GDP. There are approximately two million jobs in the region, again fairly evenly divided between its metropolitan core and the surrounding area. Its historical strength has been in manufacturing – for example, engineering and car production in the Birmingham conurbation and Coventry; and the ceramics industry in North Staffordshire, the home of such household names as Wedgwood and Royal Doulton china. Twenty-seven per cent of the workforce are employed in the manufacturing sector; this is the highest in the UK and amongst the highest in Europe. However, there are weaknesses – for

## Figure 12.1: The West Midlands Region

*Source:* © Crown Copyright Warwickshire County Council LA076880, 2000

example, labour productivity in manufacturing is the second lowest of any English region. Overall, GDP per head remains significantly below the national average (94%) despite recent growth, and standards of educational attainment are generally amongst the lowest in England. Across the region the loss of the steel-making industry and the once-strong coal mining sector left a legacy of dereliction and a weakened employment base – a major regeneration challenge.

Nevertheless, there are some encouraging signs. In recent years the West Midlands has been particularly successful in attracting foreign direct investment (FDI) in both manufacturing and service industries. For example, in 1997 over a quarter of all jobs created through FDI in the UK were in the West Midlands region. In the service sector, the regional capital Birmingham is now well established both as an international centre, with such major facilities as its International Convention Centre, and as the UK's leading financial centre outside London. Elsewhere, the region's quality environment and good communications are attracting HQ and R&D operations of prestige companies.

*Transport*

The region is traversed by a number of major motorways, including a motorway 'box' around the Birmingham conurbation. This gives excellent road access to Birmingham International Airport and the adjacent National Exhibition Centre. However, parts of the motorway network are now extremely congested, carrying in excess of 150,000 vehicles per day. A transport strategy for the region was agreed early in 1998 in which the need for selective major investment in roads, including the privately financed Birmingham Northern Relief Road, was recognised. The strategy, entitled *'Accessibility and Mobility – An Integrated Transport Action Plan'*, also sought to promote more environmentally sustainable transport policies. Thus, although pre-dating the Labour Government's 'Integrated Transport' White Paper, it foreshadowed most of the policy changes introduced nationally. Regional priorities included support for substantial investment in the main railway line across the region, which links London with the north-west of England and Scotland. However, the West Midlands was early to realise that there was still some way to go in implementing a sustainable transport policy; the private car continues to be the dominant mode of transport, carrying over 60% of journeys to work in the core metropolitan area and 75% elsewhere.

*Deprivation*

Overall, the region is one of great diversity. The metropolitan core remains more densely populated than anywhere outside Inner London (3,000 people per square kilometre) with a diverse ethnic mix, 14.6% of the population being non-white. Birmingham and the adjoining Borough of Sandwell are amongst the 10 most deprived districts in the country and there are pockets of multiple deprivation in all the districts. This is recognised nationally by the range of community regeneration initiatives operating throughout the region, including support for parts of the rural hinterland, reflecting the depth of rural disadvantage experienced in that part of the area.

Much of the region has, until recently, also benefited from its designation as an Assisted Area by the UK Government, which has enabled it to provide certain types of support to industry. However, the recent review of areas eligible for Assisted Area status has significantly reduced the extent designated for the future. The needs of the region have also been recognised by the European Union (EU). Between 1994 and 1999 the region qualified for support under two of the geographically targeted Objectives of the Structural Funds (combating industrial decline and tackling the problems of rural diversification). In addition, the region received targeted help from a number of European Community initiatives. The region will continue to receive substantial, albeit reduced, support under the new Structural Fund programmes which operate between 2000 and 2006.

**Institutional Relationships within the Region**

*Overview*

The West Midlands region contains the expected range of central and local government agencies found in other English regions. It includes the Government Office for the West Midlands (GOWM), the West Midlands Local Government Association (WMLGA), the Regional Development Agency 'Advantage West Midlands' (AWM) and the Regional Chamber.

There are also a number of separate executive and regulatory agencies of central government with a regionalised structure – notably the Highways Agency, which manages the motorway and trunk road network, and the Environment Agency, which regulates pollution and emissions on land, in rivers, and in the air. (Unfortunately, the regional boundaries for these agencies do not currently coincide with those of GOWM.)

In addition, although not directly associated with regional governance, a number of other groups of stakeholders have a key influence in the West Midlands. They include a range of voluntary, community and environmental groups; business organisations – the regional CBI, Chambers of Commerce, and so on – and the former public utilities, now privatised. The latter include, amongst others, water and electricity supplies, and especially railway companies whose commercial objectives need to be locked into the regional transport agenda.

At local level, the government structure is currently as follows:

1. Seven metropolitan boroughs, covering the core urban area of the region, the largest of which, by far, is the City of Birmingham with a population of around one million. The seven authorities are responsible for all local government functions within their areas, although some functions are still managed jointly – notably passenger transport.

2. Four surrounding county areas, within which the two-tier system of local government operates. The upper-tier County Councils have strategic planning, economic development, transport and waste disposal duties (as well as other functions – notably education) whilst twenty-four districts within those county areas have local planning and development control duties.

3. Three unitary councils have been carved out of the county areas and are all-purpose local authorities within their boundaries, with similar functions to the metropolitan boroughs.

It is, however, GOWM, WMLGA, AWM and the Regional Chamber which are the critical public sector players in shaping the spatial development of the region and the interactions between them are largely defining the agenda for debate within the parameters of Regional Planning Guidance (RPG). Each of these institutions, however, faces a period of considerable uncertainty.

*GOWM*

One of the Integrated Regional Offices (IROs) first introduced by the Government in 1994, GOWM's role has continued to evolve since that time. Most recently (in February 2000) the Government recognised, following a report from the Cabinet Office's Performance and Innovation

Unit (PIU), that different Government Departments had not been adequately integrated at regional level across England. The report's conclusions (which were accepted by Government) were that stronger and higher profile IROs were needed and that a new DETR unit be established to oversee the co-ordination and delivery of key Government policies in the regions. The full implications of this change are not yet clear, particularly with the future of directly elected regional level government still uncertain. However, it would appear that the IRO's hand as a regional player will have been strengthened whatever the future of regional governance.

*WMLGA*

Between them, the local authorities have established a West Midlands regional Local Government Association. This has full-time administrative staff and is funded by a voluntary levy on each local authority. It has created a series of indirectly elected groupings of local Councillors, embracing the full range of local government functions, which are professionally supported by senior local government officers from across the region, as part of their normal duties. Importantly for regional planning, the WMLGA's Regeneration and Environment 'Conference' is the Regional Planning Body.

The WMLGA is the latest manifestation of local government organising itself to secure co-operation and success within the region. This imperative has survived significant changes in local government's administrative structure, including the creation of three new unitary authorities. However, it currently faces two difficult tests. Firstly, there is the potential for a change in the balance of power between the political parties, particularly after the Shire County elections in May 2001. If this happens, it would be a significant shift from the near monolithic control of local government and its institutions in the West Midlands throughout the 1990s by the Labour Party. The second key challenge will be to accommodate within the ongoing regional agenda the changes that appear inevitable within most local authorities within the region. Although attention may be focused on the question of whether or not elected Mayors are the way forward for the region's cities, there are other structural issues within all authorities (for example managing the new 'overview and scrutiny' function) and in their relationships with new sub-regional quangos like Local Learning and Skills Councils. These challenges are of great significance to the regional planning function, both because of the resources and attention that their resolution will require and because they may well lead to the creation of different, competing, visions for the regional agenda.

However, there are good grounds for optimism. Most impartial commentators would recognise that local government in the West Midlands has 'had its act together' for some time. In particular, the WMLGA's predecessor body – The West Midlands Regional Forum of Local Authorities – could point to some path breaking achievements.

- It enabled the West Midlands in 1998 to be the first English region to produce a Regional Transport Strategy.
- In the last revision to RPG in 1998 it secured unanimous agreement between the region's local authorities to the regional total and distribution of new housing. Perhaps even more remarkably, the local authorities' view and that of the Government were almost identical, with only a minor difference in relation to the level of provision needed for specialised accommodation for elderly people.
- It pioneered changes in housing strategy away from 'predict and provide' towards a more sensitive approach now expressed as 'plan, monitor and manage' and recognised also
- the need for explicit recognition of affordable housing needs and market housing provision as separate, albeit related, issues.
- It secured cross-party and cross-boundary support for urban brownfield regeneration initiatives within the core metropolitan area as a key success factor for the region as a whole.

*Advantage West Midlands*

All the RDAs came into being in April 1999 with a very daunting challenge: to produce a Regional Economic Strategy (RES) at breakneck speed, whilst keeping existing programmes on the road and integrating staff from a number of different predecessor bodies. The limitations of the RESs nationally are widely documented and the West Midlands strategy is not immune from many of those criticisms; nevertheless, it does also have a number of interesting features and strengths which should provide a strong foundation for future development.

However, in the first twelve months AWM has faced some difficulty. The key problem, which led to the departure of its first Chief Executive barely a year after its creation, was its inability to develop effective partnering arrangements with its key regional stakeholders. It is likely that the problems were exacerbated by the need for key personnel within AWM to move into 'crisis management' arrangements as a result of BMW's decision to dispose of its interests in the Rover Group in the spring of 2000. However, this seemed to cut little ice with the business sector, which eventually forced the issue over AWM's relationships and its Chief Executive, although other key sectors shared the underlying concerns. As a result of these difficulties, AWM is still seen as needing to establish itself within the region.

*The Regional Chamber*

This again is one of a national network and, although not created by statute, its legitimacy is explicitly recognised by central government. In fact, the West Midlands Regional Chamber was one of the first to be so recognised, reflecting the region's degree of coherence and its ability to organise its structures. The West Midlands Chamber has sixty members – the majority drawn from local government – and is serviced by the WMLGA administration. Its role, essentially, has been to secure more inclusive ownership of the RES and to provide some degree of democratic oversight to the RDAs. However, one of the consequences of AWM's troubles has been that the Regional Chamber has not yet established an effective role for itself. This is undoubtedly exacerbated by ongoing difficulties over resourcing and the wish by local government to retain, via WMLGA, the RPB role. Given the commitment of all sectors to the Chamber, and the seniority of their representation, the present situation therefore appears to have some way to develop.

**Processes of Regional Planning**

By 1999, it was clear that a comprehensive review of RPG11, the currently approved regional plan for the West Midlands, was necessary. From the outset, the WMLGA as RPB was determined that it would adopt a new approach to its preparation. Clearly, a new approach is in part dictated by the requirements of PPG11, in which for the first time a clear steer is given on public consultation, timetabling, and so on. However, the WMLGA wanted to ensure that a new culture of active participation and partner working were embedded in its processes. In February 2000, therefore, it launched for consultation a draft project brief – essentially seeking to identify the questions to be answered by the RPG. At the launch conference a commitment was given to an 'open, transparent and inclusive process', a commitment which was strongly supported by the conference delegates.

Interestingly, requests for changes to the draft brief have been mainly minor or of a cosmetic nature. The one clear exception to this pattern was for transport, where a number of respondents highlighted the need for greater emphasis than in the draft.

However, the key elements of the RPB's commitment to 'an open, transparent and inclusive process' in practice have been three-fold: firstly, a recognition that further 'conference' style briefings and discussion are essential if the RPG is to be influenced by informed debate; secondly, explicit engagement at key stages with key regional partners; and, thirdly, the inclusion within the RPG's preparation processes of opportunities for players traditionally seen as external to the core planning process to influence the form and content of the RPG. The latter is being achieved by the creation of a series of 'Partner Reference Groups' (PRGs) which advise on a range of cross-cutting topics likely to be central to the decisions which need to be made in the RPG. Crucially, the PRGs are chaired by external partners who have no executive role in drafting the RPG. The aspiration, therefore, is that these Groups will not become bogged down in the minutiae of drafting which besets all plan-writers; instead, our partners can promote their own perspectives on each issue within a framework which is much 'looser' technically.

The reality of this approach is that some key interests are easier to engage with than others. It is no surprise, for example, that the environmental lobbyists are generally well organised and well connected – and, therefore, carry considerable influence in setting (and answering) the questions. More surprisingly, the business community, traditionally too busy 'wealth creating', is organising itself effectively to lobby for its interests, perhaps for the first time in West Midlands regional planning issues. It is too early to make a judgement on its success, but it is clearly not simply an AWM initiative; it is significantly related to the activities of a few key individuals and organisations with deeper roots in West Midlands issues.

A continuing difficulty with the approach adopted is the effective engagement of the community and voluntary sectors. These sectors are almost inevitably fragmented, concentrating as they generally do on focused and/or local issues to which engagement in a strategic regional agenda doesn't obviously offer significant added value. However, this is a challenge which regional planners must address – otherwise, their commitment to 'sustainability' is shorn of one of its central obligations – and regional planning will not be seen to be helping deliver the Government's key social inclusion agenda.

## Key Issues

*Vision*

Although the RPB's commitment to 'open, transparent and inclusive' processes are, in its view, essential to the ultimate acceptance of the new RPG within the region, it is the content of the RPG and, in particular, the resolution of a few key issues which will shape the region's destiny. At the heart of the RPG will be the RPB's 'vision' for the region. Vision statements are often (rightly) derided for their blandness and, indeed, large parts of the West Midlands vision could equally apply to other English regions. The one unique element of the vision statement is its support for 'A region with diverse and distinctive cities, towns, sub-regions and communities with Birmingham as a "world" city at its heart'. This element of the West Midlands vision identifies two key features.

Firstly, an unequivocal recognition that the region as a whole can only succeed if Birmingham, its regional capital, is seen to be a major player on an international stage. In many other UK regions it would be difficult to identify a regional capital, without a credible challenge from a rival city and, therefore, such a focused vision statement would be impossible.

Secondly, however, this does not lead to acceptance of a vision of the West Midlands as Birmingham's 'City Region'. The reality of the West Midlands is that it sees itself, to borrow a concept from the European Spatial Development Perspective, as a 'polycentric' region and it expects to develop its own spatial strategy on this basis. It will look to play to its sub-regional strengths and overcome its areas of weakness and thereby help the region as a whole climb the international league tables of performance. By way of example of the intra-regional variation, the Warwickshire/Coventry/Solihull sub-region – one million people living in the closest part of the region to Britain's booming south-east – out-performs the rest of the region economically and on some indicators betters the UK average. Statistics for the area show:

- average full-time earnings 5% above the regional average;
- GDP per head 6.5% above the regional average;
- manufacturing productivity 24% above the regional average.

Similarly, educational attainment indicators for both young people and the workforce mostly out-perform regional and national levels.

Such variations in economic indicators are important to the development of regional strategies; what looks like a stretching but attainable target for Warwickshire looks very much more difficult for Sandwell, arguably the most deprived part of our region.

*Topic Areas*

The project brief in its final form identifies forty-five issues which it is hoped the RPG will address. The issues are grouped into four topic areas.

- 'Communities for the Future', which is mainly concerned with housing issues, but also questions whether RPG has a role to play in influencing the scale and location of major education, entertainment, health, leisure and recreation facilities. The possible inter-relationship between crime reduction and regional spatial form is also being addressed.
- 'Prosperity for All', which addresses a full range of economic and business support issues, where spatial development strategy is likely to be significant. It is through this area that RPG will be expected to support, flesh out and test where appropriate the spatial dimensions of AWM's RES.
- 'The Quality of the Environment' predictably covers issues associated with the green agenda, as well as flagging up important specific and potentially controversial issues such as Green Belt policy, waste planning, minerals and energy policy.
- 'Transport and Mobility' will develop the themes of the Government's White Paper, PPG13, and the West Midlands 1998 transport strategy into a new Regional Transport Strategy in accordance with PPG11.

Although there was some debate amongst the planning professionals over the wisdom of slicing up the regional agenda in this traditional 'vertical' way, the practicalities of drafting pointed strongly towards such an approach. As a result, each topic area is being developed by a Technical Group, drawn mainly (but not exclusively) from professional local government planners from across the region, into a section of the draft RPG. Clearly, the major reservations about this approach are the risk that the 'cross-cutting' issues will not be addressed coherently and that some may not surface at all. The safeguards to prevent this include the inclusive processes described earlier, effective networking, and the personal oversight being given by local government Chief Officers and senior elected members through conventional WMLGA procedures. In addition, the draft RPG will be the subject of a full-scale, independent, 'sustainability appraisal' before it is put on deposit for formal public consultation.

*Headline Issues*

Although most, if not all, forty-five issues identified in the project brief are of significance in shaping the region's future spatial form, the 'big picture' needs to be conveyed through a much smaller number of key questions which give life to the 'vision'. The project brief has kicked this off by identifying half-a-dozen headline issues, listed below:

- Can and should growth be channelled from the areas of greatest potential to the areas of greatest need?
- Can patterns of social change which are leading to the concentration of disadvantage be effectively influenced?
- What should be the nature and form of urban renaissance and regeneration?
- How can we ensure that the distinctive needs of the Region's rural areas are met?
- What do we need to do to ensure that the right development comes forward in the right place in order to minimise the need to travel?
- Are certain areas reaching environmental limits to their capacity to accommodate further growth?

Interestingly, this statement of headline issues was not altered in the light of consultations on the draft project brief, on the basis that they were generally considered to be acceptable. However, since then they have been subject to further scrutiny by partners, as a result of which a new but related set of key decision areas (or 'strategic choices'), has been identified, listed below:

- To what extent can the RPG influence the pattern of economic growth?
- To what extent can the decentralisation of people, jobs and other activities away from the major urban areas be reversed?
- To what extent should development opportunities within the sub-regions be concentrated or dispersed?
- To what extent should significant development be concentrated in transport corridors?

These 'strategic choices' featured as the centrepiece of a second major conference in October 2000, as a prelude to the development of a more detailed draft spatial strategy early in 2001. Partners were invited to confirm whether or not these choices do encapsulate the strategic agenda for RPG development and to identify where along the spectrum of choice for each of the four areas they think the region should set its sights.

The debate around 'strategic choices' is likely to lead to further significant amendments before a consensus view of the RPG agenda can be achieved. However, if the process underway is successful, an appropriately modified form of the headline issues will set the agenda for debate at the Public Examination.

**Outlook**

The West Midlands RPB has set itself some stretching targets, covering both process and content, for the production of RPG. Its innovative approach, summarised in Figure 12.2 seeks to expose, debate and resolve differences wherever possible. The likelihood of it succeeding is not yet clear, but there are some important plus factors, not the least important element of which is a recognition that there is such an entity as the West Midlands region within which some actions need to be taken and decisions made at regional level. As a result, the 'key' regional agencies appear to be committed to the interests of the West Midlands and to working in partnership with each other, notwithstanding the stresses which threaten to pull them apart. In particular, there is a history of regional co-operation – especially, but by no means exclusively within local government – which can be seen to have brought substantial benefits in recent years. Perhaps most important is the commitment of the RPB to work in an 'open, transparent and inclusive way' and the early evidence that it is delivering on this commitment.

On the other hand, there are clearly some important challenges to be faced. Uncertainty over the future roles of the main regional agencies affects the relationships between them and, therefore, the likelihood of them achieving a broadly based consensus. Moreover, the RPG is working to a very tight schedule with a resource-hungry process which is untried locally and already running into serious difficulties nationally, especially over housing. However, the most serious challenge is over vision and policy where no amount of process will disguise underlying conflicts of interest. These may include:

- those between some environmental and economic growth lobbies;
- differences in vision between advocates of 'the city region' and the 'polycentricity' approach;
- the balance between those issues to be addressed regionally and sub-regionally via Structure Plan and UDPs.

What the RPB will need to achieve is clarity of its vision and overall strategy, whilst retaining the maximum amount of consensus over resolution of the 'headline issues'. It will also need a process for resolving outstanding differences within the region in order that they are not 'referred upwards' for decision in Whitehall. In practice, the RPB is likely to achieve a large measure of regional agreement over:

- the importance of regeneration within the metropolitan core, including support for an effective brownfield land development strategy;
- housing numbers, distribution and tenure needs.

In addition, now that the Government has signalled a major increase in resources and a shift in its stance on road construction, some degree of consensus should be achievable around a sustainable transport strategy. (There may, however, be reservations – both by the region's environmental groups if the transport strategy includes a significant level of new road construction projects and by major business interests if, in their eyes, it does not go far enough.) Unfortunately, the consensus will not necessarily increase the deliverability of the strategy. There may be more difficulty in reaching agreement over the balance of employment land release and associated investment in infrastructure between those parts of the region in most need and those with most opportunity. To some degree this reflects a north/south and urban/rural split, which will tend to focus on 'urban fringe' locations, and perhaps also on certain key market towns, and has obvious implications for the deliverability of a regeneration strategy. However, the risk to the region of not supporting an 'opportunity-led' approach is seen, by some business interests especially, as one of losing out to competing regions in Europe or North America. A similar debate may also emerge over future policy for Birmingham International Airport in the light of the impending publication of the Government's Regional Airports Strategy.

The need to resolve these matters may be an opportunity for the Regional Chamber to establish itself as the pre-eminent regional agency. With its membership substantially made up of local government representatives (two-thirds, and mostly local government's big players in the region) and its remit including an element of oversight of the RDA, it would seem to be ideally placed to mediate between conflicting interests of this nature. Only time will tell if this opportunity is given and, if so, if it is taken.

Table 12.1: New Processes of Regional Planning Guidance (RPG)

| RPG – old way | RPG – new way |
|---|---|
| • Technocratic – officer-led – policy principles and choices evolved through technical working | • Inclusive process through use of Reference Groups – policy principles and choices evolved through joint working with regional partners |
| • Local Authority politicians and Committees seen as main customer for officer working | • Recognition of much wider customer base, i.e. wider regional interest groups |
| • Reasonable level of dedicated officer support to deliver work programmes | • Major work pressures – more limited officer support but with greater administrative demands through more complex inclusive process and tight timetable |
| • Local Authority-led | • Wider ownership of the RPG process by key regional partners, including RDA |
| • Limited involvement of other interests in working arrangements, e.g. difficulty in finding appropriate organisations to represent social interests | • Direct involvement of many key regional partners through joint working practices – stimulating new regional groupings, e.g. social inclusion interests |

| | |
|---|---|
| • Generally closed processes | • More open and transparent processes structured to allow direct participation of different interests |
| • Land use focus | • New emphasis on policy integration with wider aim of providing overall spatial development strategy for the Region |
| • Dominated by housing numbers issue | • A broader agenda recognising the need for policy integration with some new approaches on key issues – e.g. greater emphasis on urban capacity/the plan, monitor, manage approach etc. |
| • Loose timetable | • Explicit target timetable (31 months) – a major challenge |
| • Consultation – through key interest groups where these existed (e.g. CBI, TUC, TECs) | • Direct participation of wide range of different regional partners and interests through Partnership Reference Groups |
| • Separate Transport Strategy | • Regional Transport Strategy as integral part of RPG underpinned by Multi-Modal studies |
| • Sustainability regarded as environmental sustainability | • Sustainability concept widened to include four key themes, including social inclusion |
| • Much description/little prescription | • Aim to be more prescriptive, harder/ sharper, regional strategy which will directly influence public investment programmes |

| | |
|---|---|
| • Sub-regions based on old concepts, e.g. Crescent Towns | • Sub-regions likely to be based on new groupings underpinned by greater in-depth analysis of the polycentric structure of the region |
| • Much repeating of Government (PPG) policy in final RPG | • Potential freedom to develop new regional approaches to national policy issues |
| • Agenda dominated by Metropolitan/City Region issues | • Greater emphasis on polycentricity including new emphasis on the rural agenda and the links between rural/urban issues |

# 13 The South West – Lessons for the Future

DON GOBBETT AND MIKE PALMER[1]

## Introduction

At the time of writing, the process of revising Regional Planning Guidance (RPG) for the South West (RPG10) was nearing completion. The Secretary of State's Proposed Changes to the draft RPG were expected imminently. After the consultation period on these Changes, the Secretary of State will publish final RPG.

This section reviews a number of the issues that arose during the preparation of the draft RPG and its subsequent stages through the Public Examination. It is divided into three sections – a brief overview of the history of the preparation of draft RPG, issues that arose (process, technical, and political), and some thoughts for the future.

## History of the Preparation of Draft RPG for the South West

Regional Planning Guidance for the South West (RPG10) was first published in July 1994. In early 1996, the Government Office for the South West (GOSW) indicated that it would like further advice from the South West Regional Planning Conference (SWRPC), with a view to publishing revised RPG in 1999.

The process of revising RPG therefore started under the previous system, where the Secretary of State was responsible for the publication of both the draft and final RPG. Consequently, SWRPC issued a 'Priority Issues Report' in December 1996, and a 'Revised Regional Strategy for the South West – Consultation Draft' in July 1998. The final version of this Strategy would have formed the advice of SWRPC to the Secretary of State.

However, following the publication of the Consultation Draft, the process changed. The new arrangements were partly formalised with the publication of Draft PPG 11 'Regional Planning' in February 1999. The Regional Planning Bodies were now required to produce draft RPG, working in partnership with the Government Offices. Also RPG had to cover a wider range of topics than previously, traditional topics needed to be covered in greater depth, and the consultations with key regional stakeholders.

These changes had clear resource implications. Up to the publication of the Draft Revised Regional Strategy in the summer of 1998, SWRPC had relied on 'voluntary' officer support from the Region's strategic planning authorities. However, the new agenda meant SWRPC had to second a team of part-time local authority staff from within the Region to complete the work. Draft RPG was produced in August 1999, although the accompanying sustainability appraisal did not appear until October.

SWRPC and GOSW commissioned the Round Table for Sustainable Development in the South West (now 'Sustainability South West') to undertake the appraisal from an independent point of view. An essential element of the appraisal was a process of continuous review and feedback so as to inform the next iteration of plan formulation. However, the period from the Easter of 1999 to the summer was one of frantic activity for the team, with weekly, and sometimes daily, changes in possible strategy and policy approaches. This made it difficult for the consultants who undertook the appraisal to keep up with progress. The final appraisal could of course only be written after draft RPG had been completed.

The Public Examination into RPG was held in March 2000, and the Panel's report was published in July. At that point, SWRPC merged with the South West Regional Chamber to form the South West Regional Assembly. Although there is some commonality with the membership of SWRPC, the Assembly embraces a wider and different membership. Whether the new Regional Planning Body will process had to be completed within a rigid timetable but involve wider follow views consistent with those of the old Body remains to be seen.

## Process Issues

*Working with the Government Office*

The Government Office was represented on the key officer groups of SWRPC throughout the production of Draft RPG. A member of GOSW was seconded to the team. Representatives also attended the elected members' Policy Advisory Group and the Conference itself. GOSW continues to be represented on Assembly officer groups, and attends key member Assembly groups and the Assembly itself.

# Figure 13.1: Key Diagram of Draft Regional Planning Guidance for the South West

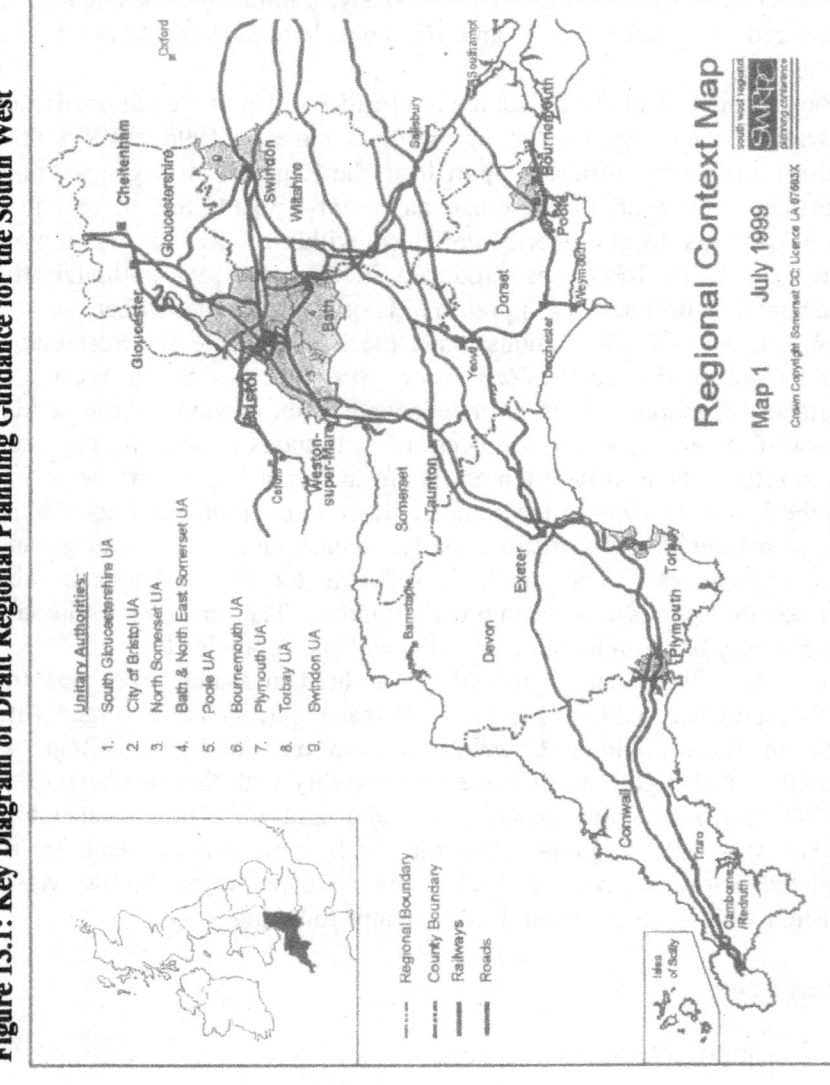

*Source*: South West Regional Planning Conference, 1999

There were both advantages and disadvantages to the close involvement of GOSW in the process. Their representatives provided some very helpful guidance during the preparation of Draft RPG. Having an official of GOSW on the team was particularly useful, not only in his GOSW role of advising on Government policy, but also because of his technical expertise. GOSW also contributed financially to a number of the consultants' background studies. However, there were times when the presence of a GOSW representative inhibited discussions, particularly when a possible strategy or policy option was known to run counter to Government policy.

This 'policing' role of GOSW may have proved counter-productive on specific matters, particularly in relation to the timetable for the production of RPG. It was clear relatively early in 1999 that some of the results of the technical research required to make soundly based decisions on the Draft RPG would not be available to meet the deadline of publication in July/August. There were problems on both sides with some of the consultants' studies. Another key element was the delayed publication of the detailed Department of the Environment, Transport and the Regions (DETR) 1996-based household projections. Without this information it was not possible to work up the housing provision figures properly. Nevertheless, GOSW were adamant that SWRPC should adhere to the timetable. References to the Government's track record in failing to publish guidance and information on time were ignored. The result was that SWRPC had to produce Draft RPG with a caveat that the housing provision would probably need to be revised before the Public Examination to take account of this additional information. Given the current significance of housing figures, this was hardly a satisfactory situation. And so it proved. Detailed DETR projections came out in the autumn of 1999, together with the results of some of the consultants' studies. Because of the short time-scale imposed on the consultants, the results of one of the studies were unsatisfactory, and further work had to be undertaken following the Public Examination. SWRPC also had to revise its housing figures, and could only publish them shortly before the Examination. This put pressure on groups represented at the Examination, and left little time for proper analysis and reflection. It also left little time for SWRPC to consider and respond to statements that were, understandably, submitted very close to the Examination.

There was frustration that GOSW was reluctant to substantiate some of their assertions, particularly where these were fundamental to policy formulation. Also, it was not always clear whether the advice given was Government policy or a personal interpretation. GOSW is based in Bristol. That area has a set of strategic planning challenges that are unique in the

South West. Yet the Bristol perspective seemed too often to colour GOSW views on the strategic approach to be taken in the rest of the Region.

One final issue is that the 'partnership' arrangement ceases following the publication of the Panel's report. This situation has the potential to create difficulties for Government Offices, particularly if the Panel suggests further technical work be done. The Regional Planning Body is the holder of all the technical data that underpins Draft RPG, and is best placed to undertake any additional technical work. However, because of the imposed curfew by the Government, its own civil servants could be left to produce Proposed Changes without the capacity to undertake any recommended further technical analysis. It would seem sensible to allow the partnership working at officer level to continue after the public examination, on the understanding that the officers of the regional planning body were to provide technical assistance only, and not lobby on behalf of the RPB.

*Working with the South West of England Regional Development Agency (SWRDA)*

The deadline for the production of the Draft Regional Economic Strategy of SWRDA was almost identical to the deadline for the production of Draft RPG. The Agency came into being in April 1999 and was required to produce a Draft Strategy by the summer. Much of the preparatory work to set up the Agency was done by a small, seconded team of officers from economic development agencies and local authorities in the South West.

Officers of SWRPC and the SWRDA team met regularly to discuss issues of mutual interest in the lead up to the production of both documents. The meetings were informal, and often took the form of brainstorming. Both sides found them valuable, as they highlighted issues that would need to be addressed both in RPG and in the Regional Economic Strategy. Both sides also agreed that the two documents should be as complementary as possible. There was some difficulty in maintaining contact during the last few frantic months before the deadlines, but the final products do bear close synergy. This was recognised and complimented on by the Panel that conducted the Public Examination of Draft RPG.

After the publication of both documents, SWRDA began to recruit staff in earnest. It began the preparation of a series of 'Action and Implementation Plans' and 'Frameworks for Action'. It also commissioned research to put flesh on its Strategy. It has not been easy to keep track of SWRDA's activities. At the same time, the Assembly was coming into being. It is still organising its member and officer arrangements, but has maintained good lines of communication with SWRDA throughout. The Chief Executive of SWRDA reports regularly to the Assembly, and there is an exchange of member and officer personnel on various sub-groups on both sides. A concordat between the two bodies is being drawn up.

*The Public Examination*

*Participants.* The participants at the Public Examination tended to be representatives of national or regional agencies that had an interest in regional planning. Professionals dominated. Local voices were rarely heard. Contributions from the heart, rather than the mind, were few and far between. While the nature of the debate was generally good humoured, it lacked passion. The exercise begged the question as to whether many residents of the South West knew, or cared, that the Examination was taking place, or, if they did, whether they would have been invited to participate. For those of us who had experienced Examinations in Public of structure plans, many of the participants were familiar. The debate had simply shifted in location and scale.

It was noticeable that many of the participants had come with the intention of objecting to the Draft RPG, but had little idea of alternative suggestions. Frequently the Panel invited such suggestions, only to be met with looks of surprise and embarrassing silence. It would be helpful if participants were asked to come with constructive ideas. Innovation should be encouraged, whether or not it accords with Government policy or best practice.

*Statements.* The Panel requested that participants' statements should be no more than two thousand words per sub-matter and should be delivered to a deadline. For the most part, participants adhered to deadlines. However, the length of some statements significantly exceeded the set limit – appendices proved a useful way of circumventing the rule. For example, some 800 pages of statements were submitted in relation to housing. To those familiar with Examinations in Public and local inquiries, this situation is nothing new. However, it is not satisfactory and unfair to all parties. Authors cannot seriously expect substantial statements to be scrutinised in detail. Other participants will feel cheated that they have had little time to consider and respond to such volumes of paper. And good luck to the Panel!

*Panel questions.* One factor that aided debate was the list of questions for discussion at each session issued by the Panel. This helped to focus the debate, although in some sessions the discussions remained general, and appeared to add little to the statements that had been submitted in advance of the Examination. Sometimes the list was too long. Further, the Panel did not always adhere to their list of questions, which was confusing.

*Publication of PPG3 and Proposed Changes to Draft RPG9*

PPG3 ('Housing') was published during the course of the Public Examination. The timing could not have been less helpful, since the Panel and participants were faced with having to interpret the new Guidance during the Examination itself.

Similarly, the Proposed Changes to RPG9 (Regional Planning Guidance for the South East) were published during the course of the Examination's discussion on housing. They were particularly relevant to the discussion on housing, since about two-thirds of the requirement for additional housing in the South West arises from net in migration, and two-thirds of that in-migration comes from London and the South East. The presence or absence in the Proposed Changes of statements concerning the desirability for the South East to 'consume its own smoke' were therefore highly relevant to the consideration of the level of provision in the South West. Again, there was little time for a proper consideration of the Proposed Changes, so that participants were invited to submit further statements following the discussion on housing.

If the Government expects the production of RPG to adhere to a strict timetable, it has to consider the effect of the publication of its own pronouncements on those timetables. The timing of the publication of PPG3 and the Proposed Changes to RPG9 did not display much evidence of 'joined-up thinking'. It is time that the Government set and adhered to a co-ordinated timetable for the publication of PPGs, White Papers and other relevant material.

**Technical Issues**

*The Meaning of Plan, Monitor and Manage*

The Public Examination debate on Plan, Monitor and Manage (PMM) was approached with some misgivings. While the phrase referred to key elements of the planning process, this 'new approach' lacked definition and expectations of it appeared unrealistically high.

Its origins can be traced to a Mr Rickett of the DETR. In giving evidence to the Select Committee on the White Paper 'Planning for Communities of the Future', he used it to explain the Government's proposal to break the mould of 'predict and provide'. For obvious reasons the conservation lobby embraced it with open arms. However, many practitioners were sceptical about the potential short-termism of this approach. SERPLAN's Public Examination highlighted a number of the development lobby's concerns about the practicalities and consequences of this new approach.

While there was no requirement to embrace PMM at the time of writing Draft RPG, SWRPC considered that the essential elements of this approach were implicit in the document in terms of (a) the framework it set for future development; (b) the monitoring arrangements that had been put in hand; and (c) the recognised need for periodic reviews.

This lack of an explicit policy on PMM raised relatively few objections at the Public Examination. However, the Sustainability Appraisal of Draft RPG was helpful in highlighting a need for an approach that is more responsive; avoids premature release of greenfield sites and prevents excessive releases that might arise from uncertainties over household formation and migration rates; and, at the same time, seeks to ensure that there is no under-provision of housing in areas of housing need.

The Panel has progressed the debate by indicating that PMM has a broad relevance to the whole of the regional planning process, both in a general sense by linking it to the Vision, Aims and Objectives, as well as in the specific context of housing. They consider it reinforces rather than weakens the importance of taking a long-term view of regional development requirements and the most sustainable way of meeting them. Indeed, the emphasis on balanced development makes the anticipation of longer-term development needs even more important. They also highlight the need for a much stronger framework for monitoring, involving objectives linked to targets and measurable indicators, as well as requiring mechanisms for regular reviews and updating of RPG.

While the debate on PMM has moved on, there are still some misgivings about its application. In the 'Plan' element, the tensions between a top-down and bottom-up approach remain. While PPG11 now accepts the need for Development Plans to consider material changes that have occurred since RPG was produced, the PPG is reluctant to concede any reconsideration of the housing provision at the structure plan stage. The process is still more akin to a 'benevolent dictatorship', with little recognition of the two-way (feedback) processes that should be involved.

With regard to monitoring, it can only be hoped that the reliance placed on it proves justified. It is certainly not an alternative to prudent long-term plan making. Nor should the investment required to collect the data be underestimated. It will rely heavily on data collected at the district level, where unfortunately monitoring has all too often proved an easy target for short-term resource savings. It is also doubtful whether many indicators will be sensitive enough to provide sufficient guidance on the need for short-term reviews. Opportunities to promote sustainable development will be lost if an over-reliance on monitoring results in the 'wait and see' scenario.

In terms of managing change, it will be interesting to see whether the Development Plan process, with its requirements for participation and inclusiveness, can ever be sufficiently responsive to 'manage' the short-term adjustments implied by this approach.

The need for a balanced approach to all these issues is important. This was recognised by the Deputy Prime Minister at the outset to this debate, when he told the Select Committee that 'I do not think planning, predict and provide is contrary to planning, monitoring and managing. One is a process and the other one is how you achieve it'. However, the debate has been useful in emphasising the importance of maintaining all these activities.

*A Sustainable National Distribution of Population and Housing*

The level of housing provision is probably the most contentious part of RPG, at least in the southern part of England. The tensions between those who believe that trend growth should form the basis of housing provision, and those who oppose such an approach for environmental and NIMBY reasons have increased substantially in recent years. The concept of sustainable development appears to have added a good deal of confusion to the debate. Both sides hijacked it to substantiate their cases. Each gives differing weight to the social, economic and environmental elements that are commonly understood to underpin sustainable development. If the concept is to become more useful then attention needs to be focused on guidance for conflict resolution.

Government guidance does nothing to clarify the position on housing provision. PPG11 sets out a number of factors that should be taken into account in deriving the overall level of provision (paragraph 5.4). These include the changing composition of households over time. Household projections, however, must not be given 'undue weight' and inter-regional migration issues 'may need to be considered'. There is no guidance as to exactly how the provision should be derived, and PPG11 now makes the situation more ambiguous than ever.

Previously, official projections of population and households were beloved by Governments (or civil servants?). They gave technical credence to the setting of housing provision in RPG and Structure Plans. They avoided the need to make difficult policy decisions, particularly where changes to past trends were desirable, but not necessarily popular. Now, at least, there is recognition that projections are one of a number of considerations that should be taken into account. However, PPG11 leaves a technical hiatus by not explaining the weight to be attached to each of the relevant factors.

One way to address this issue is to consider the distribution of population and housing on a national scale. If the concept of sustainable development is to be taken seriously, it must influence the national distribution of population and housing. It should not be relegated to more local deliberations alone. The balance between economic, social and environmental factors has a national dimension on the housing debate. It should also influence other debates, not least the aims and objectives of the Regional Development Agencies. Not every region can be the top of the European league in terms of GDP per head! At national level, the Government should set out broad aims and objectives for each region to ensure that there is complementarity between them. This agenda would affect the level of housing through its economic objectives. It would also take account of social considerations – perhaps of more weight in the north – and environmental considerations – perhaps of more relevance in the south.

There is little evidence that DETR considers all RPGs side by side to determine the cumulative effect nationally. Nor does it appear to consider what is a desirable national distribution. The present system has a tendency to perpetuate decline in areas that need growth, and growth in areas that are already overheating and have the highest concentrations of environmental designations in the country – the South West and the South East. There is a desperate need for a new national policy framework on the distribution of housing and economic activity, a 'new Barlow' (the Barlow Report of 1940 framed regional policy in the UK from the 1940s to the 1970s).

*The Sub-regional Approach*

The relative merits of a sub-regional approach, in terms of adding substance to draft RPG for the South West, have been debated throughout the process. The essential dilemmas are that the size and diversity of the region make it difficult to develop region-wide policies; and the complexity and interrelationships within the region make it difficult to subdivide the South West into meaningful sub-regions. Certainly, previous attempts to make simple distinctions between the east and west of the Region, or between urban and rural parts of the Region, had resulted in unhelpful generalisations. There was also the question of how far the analysis could go within the timetable, and the issue of the principle of subsidiarity, particularly in relation to usurping the role of structure plans.

The chosen approach was an attempt to steer a middle course through these issues. The draft RPG focused on the Principal Urban Areas, both as important parts of the regional mosaic in their own right, and as key centres for future policy-led development. It went on to set broad guidelines for the various parts of the region, while recognising that individual areas would require more locally based approaches to accommodate their individual settlement pattern.

However, it was recognised that this approach needed to be developed at a sub-regional level to consider the options for sustainable development patterns to accommodate future long-term growth, and to address specific cross border issues between authorities within the region, as well as adjoining regions. The intention was to commission key sub-regional studies to inform both the next round of structure plan reviews and the next review of RPG.

Some of the participants at the Public Examination welcomed the greater emphasis on a sub-regional approach to determining future development patterns. Others expressed concern at the possible delays in preparing structure plans while these studies are completed.

The Panel has recognised the difficulties in dividing the South West into a set of self-contained sub-regions based on major settlements. It has also endorsed the need for further studies at the sub-regional level, particularly on urban capacity and strategic options, especially where cross-boundary issues arise.

However, the Panel has also taken the view that the absence of these study findings has weakened the coherence of the strategy because they will take time to complete. In the meantime there is no clarity on the scale of development intended in different parts of the region. In the Panel's view the draft RPG spatial strategy is little more than a settlement policy, with a possible over-emphasis on the Principal Urban Areas to the detriment of the rural areas. In order to address this, the Panel has proposed an expansion of the spatial strategy, to give a stronger steer on the role of different parts of the Region, with the intention that the next review of RPG should build on this sub-regional framework.

It would seem entirely logical to develop depth to RPG through sub-regional analysis, indicating how the Region interacts and where development should go to support this process. However, the finer the policy grain, the greater the need for supporting analysis. As PPG11 recognises, the implementation of the new regional agenda will take time to mature. A stronger sub-regional approach will inevitably be an important element of future reviews, but it is also important not to go too far too fast, particularly when decisions cannot be substantiated by facts.

*Settlement Policy and the Sequential Approach*

Draft RPG guidance for accommodating future development (set within an overall context of promoting sustainable development), is to concentrate it within or close to Principal Urban Areas, on brownfield land and on land well related to transport networks and employment opportunities. The guidance for achieving this is set out in draft RPG Policies 1 and 2, and in combination they become the regional sequential approach to development. This guidance is amplified in Policies 3–19 in terms of achieving the spatial strategy and the implications for the settlement policy.

At the Public Examination there was general support for the policies of concentrating development in or close to urban areas, as well as the sequential approach for achieving it. However, there were reservations concerning the need to clarify how the draft RPG strategy was to be implemented, in particular the relative priorities for development; whether Principal Urban Areas offer the greatest potential for development, as well as a need for greater emphasis on market and coastal towns; and the need for a more interventionist approach.

The Panel had more fundamental concerns about this approach. They doubted whether the sequential approach could realistically be applied in this overarching way. In particular they considered that:

- a sequential test on its own is not particularly helpful in testing for the most sustainable form of development when there are conflicting claims of different land uses. By definition it excludes issues such as social inclusion, the need to provide for all land use requirements and creating balanced development. They point out that where the sequential approach has been used in Government policy, for retail and housing, it has been tailored to the characteristics of that particular use;

- because of the limited opportunities for brownfield development in the South West, there would be more scope for development in the Principal Urban Areas if RPG placed a greater emphasis on raising densities, promoting mixed use and proactive policies towards regeneration. Helpfully, they remind us that in order to secure more sustainable cities, housing should not always be the preferred option for previously used urban land and that development plans need to provide for all land use requirements.

These criticisms of draft RPG's approach have to be acknowledged. With the benefit of hindsight it is possible to see that SWRPC, despite misgivings, allowed itself to be unduly influenced by the GOSW view that the sequential approach was the panacea to ensuring sustainable development. The only mitigating circumstances were that the pressures to produce draft RPG within a very limited timetable had obvious implications for the time for debate, and that considerable resources were spent on trying to make the document succinct.

In looking at a way forward we need to reflect on the fundamental purpose of RPG. An essential task must be to define broad principles that assist Local Authorities and other agencies to find the most integrated solutions for sustainable development in their area and, because most situations will require a degree of trade-off between objectives, a way to resolve the inevitable conflicts. However, it also needs to be recognised that this balance between sustainability trade-offs cannot be determined at the Regional level. This is because the scope for integrated approaches is only likely to become apparent at the local level, when solutions are sought to particular problems in particular places, in the light of local needs and circumstances.

One of the lessons learnt from this particular debate is that RPG should limit itself only to establishing some of the wider principles that should influence the location of development – in essence, providing a checklist of 'Principles of Future Development'. References to the 'sequential approach' need to be seen as part of a specific checklist relating to a particular form of development, such as retail. RPG should recognise that the sequential approach will not necessarily result in the most sustainable form of development. The more important imperative is to make the best use of existing developed land to promote more sustainable, balanced urban areas that people will want to live in.

## Political Issues – the Government's Regional Agenda

The new process of producing RPG and the broadening of its remit are part of the Government's wider regional agenda. The establishment of Regional Development Agencies, the requirement for Regional Sustainability Frameworks, Regional Cultural Strategies and Regional Housing Statements are all part of this agenda. Whether or not this process will culminate in regional government remains to be seen.

The plethora of regional strategies and activity demands integration. In the South West, we believe this has been achieved reasonably well to date. Nevertheless there is a great deal of misunderstanding and confusion about the role of various regional bodies and their strategies, frameworks, action plans and so on. An integrated regional framework may help, but equally has the potential to add to the confusion.

There is some unease about whether or not the regional agenda increases or diminishes the principle of subsidiarity. The press release that announced the publication of PPG11 was headed 'Local people to lead on regional planning'. Present arrangements suggest that this is spin rather than reality.

Firstly, the role of the Regional Planning Body is to deliver *draft* RPG. After that, the Secretary of State takes over and the final version of RPG is his and his alone. If local people were taking the lead, they would be given the final say in determining the content of their own RPG. Indeed, the determination within RPG of housing provision at structure plan area level – with the expectation that this level of provision will not be altered when it comes to the preparation of structure plans – represents a withdrawal of control to the centre. Although the Secretary of State already has the power to issue directions on structure plans, the determination of the level of housing provision has largely been a matter of local determination. The situation is exacerbated by the fact that individual strategic planning authorities have no right to be represented at Public Examinations. The regional planning bodies represent them collectively, but there is little opportunity for the Panel to hear dissent from the collective line.

Secondly, the partnership working with Government Offices during the preparation of the draft, as noted above, has more than a hint of policing about it. There is strong pressure to adhere to the central policy line and a disincentive to be innovative, or make local variations or departures. We are left with the impression that regional planning bodies are simply providing the manpower to do the Government's work. Conveniently, the regional planning bodies can then be blamed for any unpopular decisions, if they are embraced in draft RPG (unless, as is the case in the South East, the Proposed Changes differ radically from the draft).

Thirdly, as noted earlier, many of the policy decisions embraced in RPG require detailed local knowledge if they are to be based on the best information. There is a limit to the amount of information a regional-level body can assemble and consider. Yet, decisions made at that level may be quite wrong in the absence of local data. The accommodation of housing provision is one example. In the South West, despite valiant efforts to assemble supply-side information, the RPB and Panel were forced to rely mostly on demand-side considerations. The tight and unrealistic timetable, now reinforced in the final version of PPG11, also limits the ability of regional planning bodies to assemble good quality information.

Fourthly, the range of subjects that RPG is now required to consider seems to be divorced from the consideration of the level at which they are most appropriately considered. PPG11 sets out tests of regional significance – not to be of national scope, to apply across regions or parts of regions and to need to be considered on a scale wider than the area of a single strategic planning authority (paragraph 1.2) – and states that RPG should not repeat national policies (paragraph 1.4). Yet PPG11 insists that a wide variety of topics should be covered – apparently irrespective of their regional significance or whether or not they pass PPG11's own regional test. We note with some amusement that PPG11 states that RPG should be concise and easily understood. PPG11 runs to 83 pages – Draft RPG10 is 57 pages long.

**Lessons for the Future**

Any regional tier of activity should help deliver public services more efficiently and with greater public ownership. Activities at regional level should be undertaken only where they could not better be undertaken at national or local level. They should be accountable to the people who live and work in the region. Wider responsibilities for regions must be accompanied by wider powers and greater resources.

At present, we have mixed feelings about the direction which regional planning is taking. There are certainly issues that should be addressed at regional level, particularly those that cover more than one structure plan area, but do not merit national attention. On the other hand, PPG11's comprehensive topic approach seems too elaborate. It could mean that a great deal of work will be undertaken that does not add value, and it could lead to resentment that the regional agenda is being imposed for political reasons rather than to provide a better public service.

We suggest that a number of issues need further thought, clarification and consideration. These are:

*Process*

- The role of Government offices should be clarified and extended. They should be given greater flexibility in the application of Government policy to take account of regional characteristics. They should overtly lobby Government on behalf of their regions.
- The timetable for the production of RPG should be more realistic. To expect RPG to be more comprehensive, to treat topics in greater depth and to adhere to an already unrealistic timescale betrays a lack of understanding of the technical and political processes that need to be undertaken to produce a sound document.
- The Government should announce and adhere to a timetable for the publication of guidance, other relevant policy documents and official projections. That timetable should take account of RPG production timetables or allow variation from those timetables where Government announcements or information will have a significant bearing on the content of RPG.
- Partnership working between Government Offices and Regional Planning Bodies should continue to the end of the process, and not terminate after receipt of the report of the Panel that conducts the Public Examination.
- Working with other regional partners should be undertaken at various levels, from formal political to informal officer. A collective understanding of the Region based on a common set of information is invaluable.
- Public Examinations should be more inclusive in terms of participants, and less dominated by professionals.
- Public Examination participants should be briefed to come prepared to make positive contributions to the debate, rather than restate objections.

*Technical*

- Regions themselves should adopt final RPG, in the context of a national plan setting out distribution of population, housing and economic activity based on sustainability principles. Barlow should be revisited.

- The scope of RPG should also be a matter of regional determination. Regions should be allowed to decide whether the topics suggested in PPG11 have a regional dimension. If not, they should be able to omit them from Draft Guidance.
- There is a need for the DETR to clarify the Plan, Monitor and Manage approach. At the present time, interpretation of this approach relies mainly on the deliberations of the Public Examination Panel. While there is a need for a more bottom-up approach to arriving at the housing figures, there should also be checks and balances to ensure sufficient homes are built.
- The principle of subsidiarity should be respected throughout. The new approach is, in some areas, removing decision-making from the local to the regional level. Yet the resources available at the regional level to undertake the work in a more effective way than at the more local level are restricted.

**Note**

[1] Don Gobbett is the County, Regional and European Strategy Manager with Dorset County Council and Mike Palmer is Team Leader (Strategic Planning) with Plymouth City Council. Both were seconded to the South West Regional Assembly to assist in the production of Regional Planning Guidance. The views expressed here are those of the authors and not necessarily of their employers or the South West Regional Assembly.

**References**

Barlow Report (1940), *Report of the Royal Commission on the Distribution of the Industrial Population*, HMSO, London.
Crow, S. and Whittaker, R. (1999), *Regional Planning Guidance for the South East of England, Public Examination May – June 1999, Report of the Panel*, Government Office for the South East, Guildford.
Crowther, I.H. and Bore, J. (2000), *Regional Planning Guidance for the South West Public Examination – Panel Report*, Government Office for the South West, Bristol.
Department of the Environment, Transport and the Regions (1998), *Planning for Sustainable Development: Towards Better Practice*, HMSO, London.
Department of the Environment, Transport and the Regions (2000a), *Planning Policy Guidance Note 3: Housing*, HMSO, London.
Department of the Environment, Transport and the Regions (2000b), *Planning Policy Guidance Note 11: Regional Planning*, HMSO, London.
House of Commons Select Committee on Environment, Transport and Regional Affairs (1998), *Minutes of Evidence (Tenth Report)*, HMSO, London.
South West Regional Planning Conference (1998), *Revised Regional Strategy for the South West – Consultation Draft*, South West Regional Planning Conference, Taunton.

South West Regional Planning Conference (1999), *Draft Regional Planning Guidance for the South West*, Government Office for the South West, Bristol.

South West Round Table for Sustainable Development (1999), *Sustainability Appraisal of Draft Regional Planning Guidance for the South West*, August 1999, Government Office for the South West, Bristol.

# 14 Regional Planning in South East England – A Case of 'Eyes Wide Shut'

CHRIS M WILLIAMS [1]

## Introduction

The South East of England is the largest and most complex of the regional planning areas. It contains a population of 18.6 million people and is the city region for London. The regional planning body for this area was The London and South East Regional Planning Conference (SERPLAN) which had been set up by the London County Council some 35 years ago to manage the relationships between London and its hinterland. Some 138 local authorities (involving County, District and Unitary Councils, and London Boroughs) made up the membership of SERPLAN's Conference.

SERPLAN was involved with planning for the region for over 30 years. However, since the last of the New Towns was designated, regional planning has been living off the capacity created by some bold strategic decisions taken at that time. As time has moved on this capacity has been eaten into until today we are faced with shoehorning more and more development into the existing settlement structure. The making of a Sustainable Development Strategy for the South East is the story of the search for a way of providing development capacity for the next 30 years within an environmentally sustainable framework.

Work on preparing a new Sustainable Strategy for the South East commenced in 1995 and a paper which the author presented to the Town and Country Planning Summer School outlined some of the key issues facing the South East and some of the key principles of the process that would be followed in drawing up a new strategy (Williams, 1995). Remarkably much of that process held together throughout the ensuing five years, despite considerable pressures on SERPLAN as an organisation. However, in this period, the technical process of producing a new regional planning strategy had to cope with an unprecedented level of uncertainty and change in the environment in which the plan-making process was being carried out. In particular, there were significant changes in Government policy, new institutions with new powers and responsibilities being set up, new guidance on the preparation of regional planning guidance (RPG) and an increasing politicisation of planning. That SERPLAN managed to produce a strategy at all was a remarkable feat and a testament to the officers and members who were involved.

## The Priorities of SERPLAN's Strategy

Over the last 30 years the South East has been characterised by massive forces of decentralisation and dispersal from all of our urban areas, not only from London, but also from many of the larger towns in the region. These forces, which have been fuelled by a heavy dependency on the private motor vehicle, has led to a break in the traditional centrifugal forces centring on London as a capital city and has given rise to much more complex patterns of movements in and around the region. This in turn has led to unsustainable patterns of development. The environmental impact and economic costs of continued dispersed growth are significant, as is the strain that further development is placing on vulnerable areas of landscape and wildlife.

Early on in its plan-making process SERPLAN produced a vision for the South East which had three key features:

- promoting an urban renaissance;
- securing a prosperous and multi-purpose countryside;
- shifting transport policy to reduce dependence on private road transport.

These three key principles were based on a view of spatial planning which was holistic and well beyond the traditional land-use planning straitjacket of former RPG documents. It was a strategy with a 'vision' which depended on it being implemented by many other regional players. It was heavily influenced by the emerging principles of the European Spatial Development Perspective. The public consultation draft and final draft were both published in 1998 (SERPLAN, 1998a and 1998b). Figure 14.1 is reproduced from the final draft strategy.

In particular, SERPLAN sought to bring about a situation where urban areas in the region were more vibrant and exciting places. It sought to do this by creating a less dispersed pattern of development, whilst still strongly protecting the countryside. It sought a better distribution of economic opportunities across the region and aimed to provide housing more closely related to economic growth. It sought more equitable access to homes, jobs and leisure, and more travel being undertaken by rail, bus, cycling and walking. Transport investment was to be better related to development, particularly through investment in passenger transport facilities. The aim was to reduce pollution overall and to protect and enhance the environment of the region.

The central tenet of the Strategy was to promote London's role as a world city. This entailed actions which helped to retain people and activities in London and to improve transport systems both within the capital and linking the capital to its hinterland. Of particular significance was the view that London should maximise its housing contribution, partly to provide increased workforce for the anticipated employment growth in the capital, but also to reduce the outward pressures on the Rest of the South East. Particular concerns for the London authorities were the emphasis to be placed on the eastern end of Thames Gateway, the links between London and the M4 corridor, key issues surrounding the Channel Tunnel Rail Link and development at Stratford, together with issues surrounding the provision of additional capacity at London's airports.

SERPLAN also identified a number of priority areas for economic regeneration. These areas included not only Thames Gateway and the South Coast Towns, but also the areas around Luton/Dunstable and the Lea Valley in the north of the region. These were identified as the areas where there would be major new development opportunities, since they were areas of substantial land and labour resources. The aim was to use major infrastructure projects to kick start regeneration in such areas and to tap into the economic spin-offs from inward investment, airports and the Channel Tunnel. They were identified as the priority areas for inward investment and footloose growth and areas where support for the development of economic clusters and skills training should be targeted.

Conversely, there were some areas of the region where an entirely different approach was to be followed. These areas, sometimes known as 'areas for economic consolidation' or 'areas of economic pressure', lie mainly in an arc to the west of London in the M40/M4/M3 wedge. In such areas a more cautious approach to further development was to be followed with employment-generating development on new sites only being allowed where it was essential to the functioning of the regional economy and was tied to that part of the region. A particular emphasis was to be on matching future job growth with the available labour supply. The removal of employment land and buildings from the existing stock, wherever possible, was to be encouraged to compensate for the fact that a very significant development pipeline existed which would further fuel housing pressures and place strains on the environment and transport networks.

**Figure 14.1: Key Diagram of Draft Regional Planning Guidance for the South East**

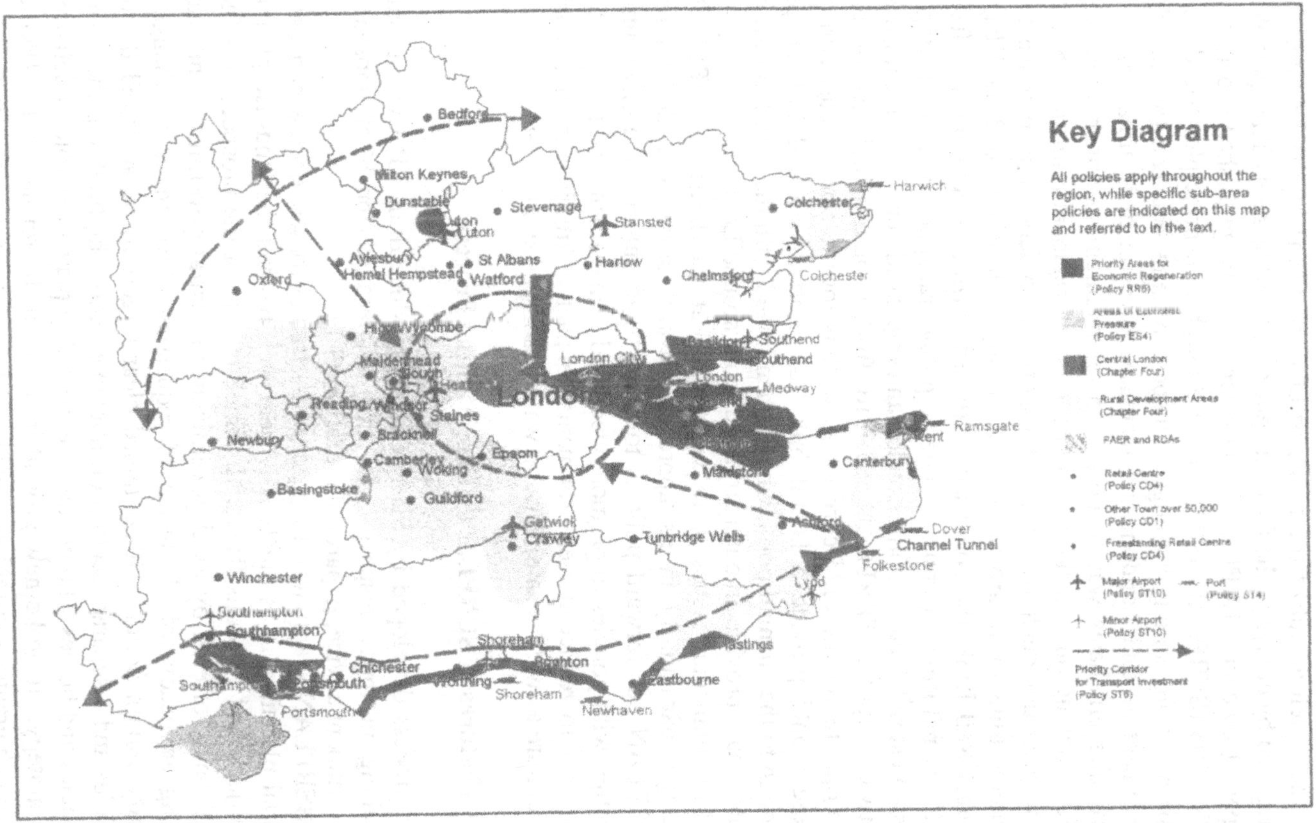

*Source:* SERPLAN, 1998

The development of the transport element of the regional strategy was hampered by delays in the Government's publication of its Transport White Paper. However, SERPLAN's Transport Strategy, published separately as SERP 503 (SERPLAN 1999) set out a number of principles based on the aims of strengthening passenger transport systems between towns, orbital movement around London, and increased investment in urban packages. Particular emphasis was to be placed on maintaining transport networks before moving on to tackling a number of key regional priorities. Some of these priorities dealt with local 'hotspots' but others were intended to be strategic links to, or around, the capital city. These included the Channel Tunnel Rail Link, the southeast-northwest corridor (including rail freight and regional freight termini), strategic public transport routes including the East-West rail route to north London and improved access to Heathrow and Gatwick airports. SERPLAN called for further multi-modal investigations into options for the M25 corridor as well as for the South Coast corridor. These would provide a lateral link from Portsmouth and Southampton through to the channel ports, which when linked with the A34 North-South route would provide an outer strategic 'box' around London.

Environmental issues were very much at the heart of SERPLAN's strategy, particularly in its approach towards sustainable development. SERPLAN put forward policies to protect and enhance the environment together with policies to promote woodland management and tree planting. It sought to reduce the environmental impact of traffic and of intensive agriculture as well as safeguarding water resources and habitats.

**Key Features of SERPLAN's Plan-Making Process**

The process developed by SERPLAN to produce draft RPG was groundbreaking in many respects. In this it presaged much of subsequent Government advice, in particular draft PPG 11.

SERPLAN's aim was to produce a strategy based on the concept of spatial planning, in as much as it went well beyond the land-use planning straitjacket of previous RPGs. It aimed to set out the prospect of a 'new deal' being offered by local government to central government and had a strong advocacy role in terms of requesting Government to change or clarify certain policies. It contained clear targets to be achieved through policies and a separate section on implementation, highlighting the various agencies and bodies who would be expected to play a key role in delivering the strategy. It broke new ground in terms of the format of the RPG and had much greater sub-regional and spatial detail than previous strategies.

Technically, the process broke new ground in a number of ways particularly in relation to its developing approach on housing. This included the mapping of all designations in the Rest of the South East on a comprehensive and consistent basis for the first time ever and the carrying out of a housing capability assessment across the whole region at a district by district level. A substantial amount of effort went into producing a technical and policy challenge to the Government's 1992-based household projections, which identified the considerable range of uncertainty associated with such forecasts. This led SERPLAN to develop what the Government now calls the 'plan, monitor and manage' approach to housing.

Perhaps the most significant features of the process were that it had at its heart an environmental assessment as an integral part of plan making rather than as a bolt-on extra, and that it sought to be as inclusive as possible and involved key stakeholders in plan preparation at various stages. Representatives of the HBF, CPRE, the Government Offices, the Environment Agency, academics and many other partner organisations were involved in different aspects of the technical work. It also involved wide ranging consultation, not only carried out by SERPLAN itself, but also arranged by member authorities in local communities throughout the region. This included the circulation of leaflets by member authorities explaining the draft Strategy and there were public meetings in certain key areas. The public meeting in East Sussex even included representatives from the Dieppe Chamber of Commerce.

## The Relationship with the Regional Economic Strategy

SERPLAN was fortunate in as much as it had produced and submitted the draft RPG to Government before the South East of England Development Agency (SEEDA) was established. SEEDA was drawing up its Regional Economic Strategy (RES) at the time of the Public Examination into the draft RPG. Prior to this there had been a joint meeting of SERPLAN's Members Policy Group with the Chairmen Designate of the three Regional Development Agencies (RDAs) that covered London, East of England and the South East. That meeting led to the identification of some 'quick wins' that could be worked upon jointly between SERPLAN and the RDAs. Amongst the areas for joint co-operation were further work on Thames Gateway, the East-West Rail Link, the South Coast road and rail links and access to the region's airports. SEEDA also had the benefit of SERPLAN's regional analysis and of its spatial strategy. It is not surprising, therefore, that the RES accepts the broad structure of SERPLAN's strategy, including the priority areas for economic regeneration, as a starting point together with the key regional strategic transport links. The differences between the RES and the draft RPG are matters of degree rather than substance, with the RES being rather more bullish about allowing economic growth in the economic pressure areas to the west of London and not giving as high a priority as SERPLAN would like to attracting inward investment to the economic regeneration areas.

## The Crow Report

A Public Examination into SERPLAN's proposals was held in Canary Wharf for six weeks during the summer of 1999 under the chairmanship of Stephen Crow, as a means of exploring the issues surrounding RPG. This process left a lot to be desired. The participants were allowed to make unsubstantiated assertions about issues and opportunities in the region and there was little opportunity for debate or discussion. Whilst SERPLAN's team produced masses of evidence to back up its case, other participants were remarkably shallow. The process resulted in a report which became known as the 'Crow Report' (Crow, S., and Whittaker, R., 1999).

This took as its starting point the view that the South East was the engine for economic growth in the national economy. Consequently, it was in the national interest for its economy to be further developed throughout the region. Growth was to be steered to areas of economic regeneration and to four major areas of plan-led expansion (APLEs) at Milton Keynes, Stansted, Crawley-Gatwick and Ashford. Perhaps most controversially the Panel recommended a level of housing provision based on the principle of providing a home for all who live, or come to live, in the region. This was to encourage planning for a level of housing development that would meet all demands. It became known as 'predict and provide plus'. In addition, it recommended that the Green Belt boundary should be rolled back in areas of strong development pressure and that constraints should not be allowed to restrict potential for economic growth at places such as Ashford.

SERPLAN challenged the Panel's report on four important elements where it saw a conflict with Government policy:

- that the Crow Report did not accept the basis of sustainable development;
- it undermined the Government's attempts to bring about an urban renaissance;
- it did not accept demand management of traffic on the road network;
- it completely disregarded the democratically arrived at objectives in the SERPLAN strategy and sought to put in place the Panel's own objectives.

SERPLAN's view was that the Panel's approach would stifle economic growth in the buoyant areas, as the resources to fund the necessary infrastructure and services failed to keep pace with development. More congestion in such areas would lead to increased pollution and environmental deterioration which would have damaging effects on business. The early release of greenfield sites and a diversion of resources could undermine the policies of urban renaissance. The economic potential of the region would be reduced due to the failure to make effective use of labour resources in parts of the region with high unemployment. It is not surprising that there was such a tremendous public outcry about the Crow Report and the Government distanced itself from its recommendation.

**The Housing Debate**

When SERPLAN first embarked on the preparation of RPG the 1992 household projections had just been published suggesting 4.4 million new households in England and Wales between 1991 and 2016. It is not surprising that the question of 'How many houses shall we plan for and where do we put them?' should completely dominate the regional planning process throughout this period. However, regional planning is about far more than just the number of houses that should be planned for and yet this question continued to dog the attempts to produce a sustainable development strategy for the South East.

At the outset of the process SERPLAN held a Members' seminar on the household projections, inviting various key speakers to examine the forecasts and some of the trends on which they were based. SERPLAN's Demography Subgroup had been tasked with carrying out a critical appraisal of the projections and advising on their veracity. In parallel with this, member authorities carried out an exercise to map all planning designations throughout ROSE and undertook a capability study to assess the scope for accommodating further development. Joint meetings were held between representatives of the DoE, HBF, CPRE and SERPLAN to debate some of the key issues surrounding the household projections and their implications for RPG. In addition, SERPLAN commissioned a number of sub-regional studies from its member authorities to look into particular areas and a separate group was set up to investigate the scope for new and expended settlements.

From the very outset SERPLAN member authorities were concerned about the level of housing provision implied by the household projections. The population of the region was projected to increase by 9%, but the total number of households would increase by 25%. This equated to an increase of 1.73 million households in the South East region, with 1.1 million in

ROSE. SERPLAN also estimated that on the basis that at least some 40% of all new households would find it difficult to gain access to the private housing market, a public subsidy on a massive scale would be needed. It also raised questions as to whether economic growth could support household growth throughout the region and conversely what level of housing was necessary to support the economy.

Member authorities were clear in their desire to develop a new approach, which moved away from the traditional 'predict and provide'. SERPLAN therefore proposed a battery of economic instruments and changes in social policy, which would help address the housing issue. It clearly set out the need for new measures to minimise the number of empty homes and for additional powers for local authorities to control the range and type of dwellings that could be permitted through the planning system. It also advocated an adjustment in planning policies to increase the number of affordable houses, and extra resources to local authorities and housing associations to provide directly more affordable homes. It set out clearly the need for fiscal changes affecting refurbishment and conversion (for example zero rating VAT) as well as fiscal changes to discourage greenfield and encourage brownfield development. Finally it set out the need for measures to encourage sharing and incentives to encourage single persons to relinquish the use of large houses.

As the technical work progressed, the political debate was engaged within SERPLAN's Members' Policy Group and the Conference itself. At each successive stage Members rejected options involving both higher levels of housing development and new settlements. The Conference further toned down references in the consultation document to the need for new settlements in the longer term. Nevertheless SERPLAN was able to publish a public consultation document (SERP400) in which three options for housing amounts and distributions were articulated (SERPLAN, 1998a).

Although wide-ranging consultations were carried out on the whole of the draft RPG, inevitably the bulk of the response focused on the question of housing. As a result of the public consultation response SERPLAN commissioned further work from the London Research Centre to produce revised household projections. The local authorities were asked to review the level of provision that they could accommodate in each area which resulted in even lower levels of provision being put forward. At this stage also the results of London's housing capacity study became available and confirmed that London could accommodate its projected number of households without any further pressure being placed on ROSE.

SERPLAN's approach to 'plan, monitor and manage' was significantly different from that of the Government. The approach, which finally emerged from the techno-political process, was one which committed local authorities in the region to a base level of provision of 860,000 additional dwellings outside London. It then encouraged all member authorities to carry out urban capacity studies on a consistent and thorough basis, to bring forward additional land within urban areas and on previously used land. To take the provision up to around 914,000 additional dwellings (which was based on meeting the LRC projected increases) some of this would also include some releases of greenfield sites following the application of the sequential approach through Structure Plans. Finally, if it was necessary to go beyond 914,000 houses, a series of new settlements were proposed – the criteria for which were clearly articulated in SERP500. By this method, SERPLAN would manage upwards the level of housing provision, culminating in a programme of major new towns which would set the framework for development for the longer-term beyond 2016.

In implementing this approach, it was intended that SERPLAN would develop a new role. Whilst all local authorities were to carry out urban capacity studies to bring forward the optimum number of units in urban areas, they were also to start the reviews of all Structure Plans and Unitary Development Plans immediately. SERPLAN advocated that these should be adopted within 30 months following the approval of the RPG and all local plans approved within 54 months. This approach shifted the responsibility for determining the amount and distribution of housing away from the RPG process and into the Structure Plan/UDP process. It would require greater involvement by SERPLAN in policing RPG and taking an active part in the Structure Plan Examinations in Public around the region. SERPLAN even floated the idea that it should be given new statutory responsibilities comparable to those given to the London Planning Advisory Committee (LPAC). Clear criteria were set out in SERP500 which would allow SERPLAN (and the Secretary of State) to assess whether individual local authorities were bringing forward sufficient land to meet their regional share.

Following the publication of the Crow report, the Government finally published draft RPG for consultation in March 2000 (Government Office for the South East, 2000a). In this the Government rejected both SERPLAN's original draft RPG (SERP500) and the recommendations of the Crow Report. However, there were five things wrong with the Government's own approach:

- It did not follow the advice in draft PPG11. Regional Planning Bodies were to be given the responsibility for producing RPG to ensure local ownership. SERP500 was locally-owned and had developed an approach to self-enforcement and monitoring of Structure Plans, together with proposals for speeding up the

- production of Structure Plans and Local Plans by setting target dates.
- The overall level of housing provision proposed by Government of 43,000 houses per annum for ROSE 1996–2016 was considered to be too high and not achievable. The first five years of this period had already gone and average completions achieved were running at 39,000 per annum. The plans for the next five years (2001–2006) were mostly set and although it might be possible to increase provision a little through urban capacity studies, certainly there was no significant additional contribution likely to be achieved up to 2006. This would then result in a requirement for house building to average 47,000 dwellings per annum 2006–2016. Such a rate of house building has never been maintained in ROSE.
- The distribution of housing did not reflect the spatial strategy accepted by both Government and SEEDA. The baseline figure in SERP500 was only ever intended as a starting point. SERPLAN recognised that there were major technical inconsistencies in the baseline figures for individual counties and had intended that these be ironed out in the next phase of the work. The Government's proposed distribution provided a 'double whammy' for those authorities who were more advanced in their Structure Plan preparation and had carried out comprehensive urban capacity studies. 'To them that hath made shall be given even more.' Nor did the distribution reflect the economic re-balancing of the region which was a fundamental part of the overall strategy.
- No additional mechanisms were suggested for tackling serious housing issues, either for bringing forward development on previously developed land, or powers to control the mix of house sizes and types. Nor were there any additional powers and resources to deliver a step change in the provision of affordable housing. All of these were seen as prerequisites to achieve even the level of housing that SERPLAN was contemplating.
- The approach advocated by Government was not 'plan, monitor and manage'. A fundamental difference between Government and SERPLAN is that SERPLAN's approach is one of setting a baseline provision figure and managing up as needed, whereas the Government approach is to set a high target figure and manage down. The Government considered that housing could be managed down from a high target figure if environmental impacts were too great or if infrastructure was not being provided in time or resources were not available for affordable housing. But it did not give further strengthening to policies on phasing the release of land and crucially

there was to be no local testing of the level of housing provision through Structure and Unitary Plans. This was a key difference from previous distribution exercises since RPG would set down the level of provision to be implemented through Statutory development plans, without giving individual authorities the ability to 'test' the level and distribution through the Structure Plan process. This removal of the 'safety value' inevitably affected the behaviour of some member authorities.

I took the view in advising SERPLAN's Conference that the consequences of the Government continuing to insist on this overall approach would be that it would not achieve its objectives of urban renaissance and sustainable development nor the target of 60% of development on previously-used land. It is hardly surprising, therefore, that local government in the South-East remained to be convinced that this was the best way forward for the region as a whole. The Government would continue to have problems with Structure, Unitary and Local Plan Inquiries and the consequences of the Government trying to force high levels of development on local authorities would mean continuing delay in the review and updating of development plan system.

**Regional Planning Guidance Finally Published**

The Draft Revised RPG was finally published in December 2000, one day before SERPLAN's final Conference (SERPLAN, 2000b). The date of the Conference had been fixed several months earlier in anticipation that the RPG would be published in early November. This would have meant that there would be adequate time for an analysis to be presented to the Conference before it finally wound itself up. In the event, this was not to be. A cynical person may well say that the delay in publication was deliberate!

This version of the RPG had changed yet again. Thankfully the Government (this time DETR Headquarters rather than the Government Offices) had reverted to the more holistic and integrated approach advocated by SERPLAN and, as a consequence, the document is much closer to being a spatial development strategy. Many of the suggestions put forward by SERPLAN in terms of creating an urban renaissance, the spatial structure of the region, the approach towards transport policies and protection of the environment had been incorporated. Most critically Government accepted an approach to housing which is remarkably similar to SERPLAN's approach to 'plan, monitor and manage'. It identified a level of development (at 39,000 dwellings per annum for the period up to

2006) which is roughly comparable to the level of development currently provided for in approved Structure Plans, and encouraged local authorities to carry out urban capacity studies as a matter of urgency to bring forward further land. It also continued to press ahead with the studies for development capacity in the major development areas of Ashford, the M11 corridor and Milton Keynes, which will feed into the next round of RPG. The only major concern in relation to the overall level of development is a comment that housing provision might have to rise to around 43,000 dwellings a year in ROSE after 2006. The local authorities considered that this would prejudge the outcome of the urban capacity studies and would be an uneasy compromise between SERPLAN's bottom-up approach and the Government's insistence on high housing targets.

Because Government had not been able to produce a housing distribution on a county by county basis that reflected the regional strategy, it reverted to the building rates contained in the previous RPG9 produced in 1994. Whilst RPG9 did attempt to shift development pressures from the west side of the region to the east, it was not an accurate reflection of the more finely grained strategy of SERP500. In examining the distribution, it was clear that some counties found the allocations acceptable, particularly those who had identified their urban capacity as part of the baseline for SERP500. Those who had not identified their capacity saw large increases in the housing requirement that they were expected to find.

At the final Conference political opinion was divided as to whether to accept the overall level of development or whether to continue to object to Government. In the end Lord Caernarvon, as Chairman of Conference, persuaded the members that since opinion was so divided, it would be better for SERPLAN not to comment on the document. That was where it ended. SERPLAN, having spent five years debating the regional strategy, failed to comment on the final version!

**Politicisation of SERPLAN and the Regional Planning Process**

The production of RPG9 was an altogether simpler process. At that time, county councils and LPAC were the strategic planning authorities in the region. Planning had only just become a major political issue at the local level and the production of RPG was largely regarded as a technical process. Indeed, the county by county allocation of housing was agreed 'in a smoke filled room' between the County Planning Officers and the Chief Planner for LPAC and then ratified by conference.

By 1995 when the technical work on the strategy started SERPLAN had become a very different organisation. District Councils became much more active in SERPLAN and secured member-level representation. This was as a prelude to the debates around local government reorganisation. Whilst colleagues from the District Councils made valuable technical contributions and gave insights into how policies would operate at the local level, there continued to be an air of county/district rivalry. This, coupled with an increasing political concern about levels of housing by authorities who had to deal with individual planning applications, raised the stakes in SERPLAN's debate about housing. Following local government reorganisation, the unitary authorities became a major political force within SERPLAN. They were mostly Labour controlled and, by and large, supported Government policies. The unitaries from Berkshire posed a different set of issues since, although they were unitary authorities in their own right and of different political persuasions, they were also members of the Joint Strategic Planning Unit for Berkshire and, wherever possible, attempted to act collectively (as though they were a county!).

Over the five years there was a significant change in the relationship between London and the ROSE authorities. At the start of the process London was very active and fully signed up to joint working within SERPLAN. A number of factors affected that relationship over time including a change in the financial relationship between London and ROSE and the increasingly inward focus of London politicians prior to the establishment of the Greater London Authority (GLA) and the election of the Mayor of London. LPAC seemed to disengage with SERPLAN and instead focused on the preparatory work for the Mayor's Spatial Development Strategy, largely only participating in the work of SERPLAN to protect London's interests. Nevertheless, LPAC was a sizeable voting block within the SERPLAN Conference and invariably voted for higher levels of housing development in ROSE than the majority of ROSE authorities.

Over the period SERPLAN became increasingly prone to party politics. In the latter years the counties were almost exclusively Conservative – controlled and were consistently arguing for lower levels of housing development. They were often supported by the Liberal Democrats from some of the District Councils who were concerned about the local impact of housing development. On the other hand, the Labour authorities from the larger urban areas tended to argue for higher levels of development on the basis that this would provide a better opportunity for meeting housing needs and would support economic growth.

Before each meeting of Conference and the Members Policy Group there would be meetings of the political groups as well as separate meetings of the district councils and the unitary councils. It therefore became very difficult to predict how particular issues would be resolved. Conference would sometimes split on party political lines, sometimes on county/district lines and sometimes on geographical lines, depending on the type of issue. With such an overtly political process, the role of officers became very tricky. Whilst all officers who contributed to the SERPLAN process attempted to act in the best interests of the region as a whole, inevitably they were there as representatives of their own authority. In some cases, therefore, the debates within the technical officer groups were a forerunner of the debates within the political meetings of Conference.

I was particularly fortunate in that Buckinghamshire County Council (and before that East Sussex County Council) allowed me to operate in a non-partisan way (other colleagues represented the County Council's point of view on the officers group). I was therefore free to act objectively in the best interests of the region. On a number of occasions I had to address the Conference giving what I knew would be unpalatable professional advice. On these occasions, when I was making 'career limiting' presentations I always acknowledged that it was for Members to make decisions, not only in the best interests of the region as a whole, but also to look after the interests of their own particular areas and be guided by their own political groupings. My role, as the lead technical officer, was to provide them with advice on the consequences of the options open to them. They were always free to set aside my advice (and often did!). It was clear from the technical advice that the RPG should contain a mechanism for providing for a higher level of housing development, which met local needs around the region. SERPLAN's approach towards 'plan, monitor and manage', leading to a programme of new settlements, would have provided such a growth. However, the politics of SERPLAN and the politics between local government and central government got in the way. It is a matter of regret, but completely understandable, that Members chose to set aside that advice with their eyes wide shut. Given this level of politicisation and the prospect of the general election it was inevitable that the RPG would become a political football within SERPLAN with what appeared to be less and less weight put on the technical work to support decisions taken. It was also not surprising that later stages resembled a game of political poker with SERPLAN pitching its opening 'offer' at a level lower than it was prepared to accept and the Government pitching its opening offer higher than it would eventually accept. The end result was probably a lower level of housing development than would otherwise have been the case if SERPLAN had put forward a 'reasonable' level at the outset.

**Disappointments**

SERPLAN can rightly be proud of its achievements. It had been at the leading edge of regional planning for the last 30 years and its work developed a number of important concepts that are now firmly embedded in Government policies. These included the approach towards urban renaissance, the holistic spatial development strategy, a more integrated approach towards transport, and a method of implementing 'plan, monitor and manage'. It has developed a strategy which addresses the key problems in the region and has developed a spatial structure of the region which has been embraced by Government and SEEDA. It has a formidable record of producing high quality technical work to support its views.

Having said that I had a number of disappointments. Some of these are disappointments with SERPLAN itself:

- SERP500, whilst a remarkable document, would certainly have benefited considerably from more time to edit it and simplify it. Because of the timetable that we were operating under, and the need to try to reflect a consensus of 138 different authorities, SERP500 is not an 'easy read'. This means that some of the clarity of thought is lost in the document itself.
- I was disappointed that SERPLAN ended with such limited ambitions for the region. Not only did this end up with too much lowest common denominator planning but it shied away from taking some bold decisions. My belief was that we had got to the point where we needed to break out of the existing settlement pattern and establish a spatial framework for the next 50 years that would benefit successive Regional Planning Bodies in the same way that SERPLAN has benefited from bold strategic decisions taken 30 years ago. There is a limit to how much further development can be accommodated within the existing settlement pattern and, in the longer term, we need major new development areas to accommodate further growth around the region. Whilst SERPLAN embraced the concept, it was not able to identify the specific locations publicly, although a substantial amount of technical work was done to identify potential areas.
- My major disappointment is with Government. The production of the RPG was a golden opportunity to produce regional strategy as a partnership between local and central government. I found the involvement of Government officials throughout the process distinctly unhelpful. At a time when they could have advised on how Government policy was changing (for example the development of the Transport White Paper) there was complete silence; and in the housing debate, instead of working jointly with SERPLAN to find a way forward, Government officials insisted on an approach which could be described as 'plan high and manage

down'. Thankfully those who wrote the final version of the strategy were much more enlightened about the approach.

- I was disappointed with the process itself, which effectively prevented any ongoing technical discussion between Regional Planning Body and Government from the time that the draft RPG was submitted. The final version of the RPG would have benefited from further discussion and debate about the practicalities and implementation of some of the policies. Because this is a quasi-judicial stage Government officials are debarred from talking to the RPBs.
- I was also disappointed that the advocacy for new powers to deliver more affordable homes had not been fulfilled by Government, but there were already signs of major shortages of affordable housing, particularly for key workers. Many of the powers suggested in SERP500 would have helped to deliver a greater proportion of affordable housing in the medium term.
- Perhaps my greatest disappointment was at the lack of trust that central government showed to local government and that production of the RPG was a huge lost opportunity. SERPLAN offered a radically new approach towards not just regional planning guidance but also the whole cascade of planning advice through the development plan system. SERPLAN, as a wholly owned subsidiary of local government, was able to draw up a strategy which achieved a consensus and would have then policed it through the Structure Plan and UDP system. It offered the prospect of, not only a strategy which was more finely tuned to the needs of the region, but also delivered more up to date plans. What we were left with instead was a level of development which was unacceptable to many authorities, and the prospect of 'hand-to-hand combat at every structure plan EIP in the South East for the next five years'.

## Note

[1] Chris M. Williams is now the Chief Officer of Buckinghamshire County Council. He was Chairman of SERPLAN's Strategy Co-ordination Group and lead technical officers. The views expressed in this article are personal and do not reflect the views of either SERPLAN or Buckinghamshire County Council.

## References

Crow, S., and Whittaker, R. (1999), *Regional Planning Guidance for the South East of England: Public Examination May-June 1999. Report of the Panel*, GOSE, Guildford, October 1999.

Government Office for the South East (2000a), *Proposed Changes to Draft RPG9 and reasons for the Proposed Changes*, GOSE, Guildford, March 2000.

Government Office for the South East (2000b), *Draft Revised Regional Planning Guidance for the South East (RPG9)*, GOSE, Guildford, December 2000.

SERPLAN (1998a), *A Sustainable Development Strategy for the South East, Public Consultation* (SERP 400) May 1998.

SERPLAN (1998b), *A Sustainable Development Strategy for the South East* (SERP 500), SERPLAN, London, December 1998.

SERPLAN (1999), *SERPLAN's Additional Transport Submission to the Public Examination Panel* (SERP 503), SERPLAN, London, April 1999.

SERPLAN (2000), *Draft RPG9 – The Government's Proposed Changes: The SERPLAN Response* (SERP 600), SERPLAN, London, June 2000.

Williams, C. (1995), 'Into the next millennium: the future of regional and strategic planning', Proceedings of the Town and Country Summer School 1995.

Williams, C. (1999), 'Breaking New Ground', *Municipal Journal*, vol. 9, April 1999.

# PART V: CONCLUSIONS

# 15 Some Interim Conclusions: Regional Planning Guidance and Regional Governance for the 21$^{st}$ Century

JOHN GLASSON

## Introduction

Regional planning – taken in this chapter as regional strategic planning, planning for a region (i.e. intra-regional not inter-regional) – has had a roller coaster of a ride over the last 40 years. In the heady days of the late 1960s and early 1970s, we had the Regional Economic Planning Councils and Boards and some very innovative planning, including the Strategic Plan for the South East, the Strategic Plan for the North West, and the Northern Region Strategic Plan. Only 10 years later in the mid-1980s, Breheny and Hall (1984) were writing of 'the strange death of strategic planning' and the victory of the 'know-nothing school'. There was a gradual revival in the early/mid-1990s, with the creation of Government Offices for the regions, and the introduction of a skeletal form of Regional Planning Guidance (RPG). But post-1997, there has been a plethora of activity, with regional planning and governance going through possibly its most intensive period of activity for several decades.

It is hard not to be optimistic. But does this plethora of activity mean real progress – at its simplest, is the new system working, is it successful? Or is it a case of new initiatives, but old problems, and even more plans and policies to collect dust on the shelves? This of course begs the question of how to judge success. What is the purpose of regional planning? At its most basic, regional planning is an exercise in persuasion, seeking to encourage those agencies with the power to act and manage regional development to adopt and use agreed strategies and to follow particular guidance in the interests of achieving identified goals or vision and consensus on 'net regional benefit'. More specifically, it is a response to contemporary regional issues – metropolitan growth, rural/industrial decline or underdevelopment, regional imbalance. Wannop notes (1995) there is a strong regional planning imperative, because regional planning issues endure.

Regional planning has not been without its successes; many of these relate to the containment and management of metropolitan growth. But the activity is fraught with many problems. It is often seen as the cuckoo in the nest between local and national levels. It is often more politically dependent than most forms of planning, often lacking the power base and legitimacy of an underpinning level of government. It suffers from political mistrust, for 'empowered' regions can be a significant force in the country (note the extremes to which the UK Government went to influence who became Mayor of London in 2000 – if London can be seen as a region). It may also suffer from conflict between physical/land use planning and economic development planning, between intra-regional and inter-regional planning, and between the many regional planning and development stakeholders. Many regional planning exercises have also placed too much emphasis on regional analysis and strategy formulation, and not enough on how to implement the plan/strategy. Martins (1986) suggests that effective regional planning needs (a) a sponsoring organisation, with an appraisal and approval network (i.e. for strategy formulation), (b) an intelligence network (for communication and co-ordination) and (c) an influence network (for co-ordination and control). In particular, if regional planning is to successfully cross the formulation-implementation gap, it needs continuity, co-ordination (to integrate an array of agencies and activities) and control (the power to make things happen). So, against such considerations, how is the new system doing?

**Researching a System in Flux**

The information base for this chapter comes in particular from two ESRC supported activities. Firstly, it draws on an ESRC Sponsored Seminar Series on Regional Planning and Governance in the English Regions, which during 1999 and 2000 brought together academics and practitioners primarily from the South East, South West, East Midlands and West Midlands regions to map out, discuss, and seek to explain the changing 'landscape'. The seminar focused on topics such as the Regional Public Examination, the relationship between RPG and the Regional Economic Strategies of the RDAs, and the changing institutional environment of regional planning. At the same time, the paper draws on a current ESRC research project on Regional Planning and Governance in England, which addressed two major research questions:

- How far do the new governance forms created in the 1990s and beyond generate a renewed regional planning process and content?
- Will this new institutional and policy context allow or force a new regionally diverse planning system, with genuinely divergent processes and content in different sorts of regions?

As such, the ESRC research project seeks to clarify how responsive to regional issues can the system be. How much divergence is there room for in a seemingly converging system? The aim is to build a model of the way the new planning forms are evolving, given the quite novel relationships which planners are having to form with other actors. These forms include those of the early 1990s, such as Government Offices for the Regions, and of the post-1997 period, such as Regional Development Agencies (RDAs), Regional Chambers/Assemblies, revised formulas of RPG and the revisions to PPG11 (DETR, 2000). This new context is especially evident in the economic and business field being articulated within the RDAs, through their Regional Economic Strategies, but it is not limited to that. For example, English Nature has recently completed a set of regional studies. Figure 15.1 provides an overview of the evolving regional planning and governance environment, 'governance' being seen as the tendency for a broadening-out of policy-making processes in many fields in recent years. The Government's Performance and Innovation Unit (PIU, 2000) identified 43 strategic plans that local authorities have to prepare, and others, such as Regional Cultural Strategies, are imminent.

The focus here is primarily on the regional planning and development agencies. For example, for the South East, these include: the Government Office (GOSE), the local authority regional organisation (SERPLAN) soon to be replaced by the regional planning element of the South East of England Regional Assembly (SEERA), and the Regional Development Agency (SEEDA). The relevant documentation output includes: RPG – RPG9 for the South East (DETR, 2000); Regional Economic Strategies (SEEDA, 1999), and also Regional Transport Strategies, and Sustainability Frameworks.

**The 'New' Regional Planning Guidance**

The new RPG emanated from a DETR Consultation Paper 'The Future of Regional Planning Guidance' (DETR, 1998) and draft Planning Policy Guidance note, PPG11 (DETR, 1999). The new round of RPG activity replaces the regional guidance notes produced earlier in the 1990s. Two key points in the new approach are:

- the need for RPG to deal with a wider range of issues than it traditionally has (i.e. wider than land use and transportation); and
- the need for wider involvement and participation in the preparation of Guidance within the regions.

The regional planning guidance is produced by local authority groupings (such as SERPLAN, or in the East Midlands, the East Midlands Regional Local Government Association (EMRLGA)). These 'conference' arrangements include Members' groups comprised of local politicians, and officers seconded from local government, plus normally at least one officer from the relevant Government Office. Participants or stakeholders are involved in the process through focus groups or other participatory mechanisms. Stakeholders in most regions are selected from a series of regionally structured environmental, business and other interest groups and quasi-government agencies. Important players in most regions include the Confederation of British Industry, Friends of the Earth, Council for the Protection of Rural England, House Builders' Federation, and the Environment Agency. The draft RPGs are produced in accordance with the requirements of the Government Office in each region. Ultimate responsibility for the RPGs lies with the Secretary of State, following a Public Examination.

The 'Regional Guidance for the Spatial Development of the East Midlands' (EMRLGA, 1998) saw its role as follows: 'to set out an integrated spatial development strategy which encompasses proposals for the development of the region's economy, its infrastructure, its housing and other land use needs, and proposals for the conservation and enhancement of the natural and cultural environment for the benefit of all the region's citizens; to incorporate the key elements of the Regional Transport Strategy (the full strategy will be available in a separate document); to set the spatial development strategy within the context of moving towards more environmentally sustainable living patterns; to involve all the region's stakeholders in a debate about the future directions of the region; to provide a framework within which Structure Plans and other more specific strategies and programmes can be prepared on a consistent basis; and to be developed in a complementary way alongside the Integrated Regional Strategy (IRS) and the RDA's regional economic strategy'.

Figure 15.1: Network of Principal Government Agencies at National, Regional and Local Level

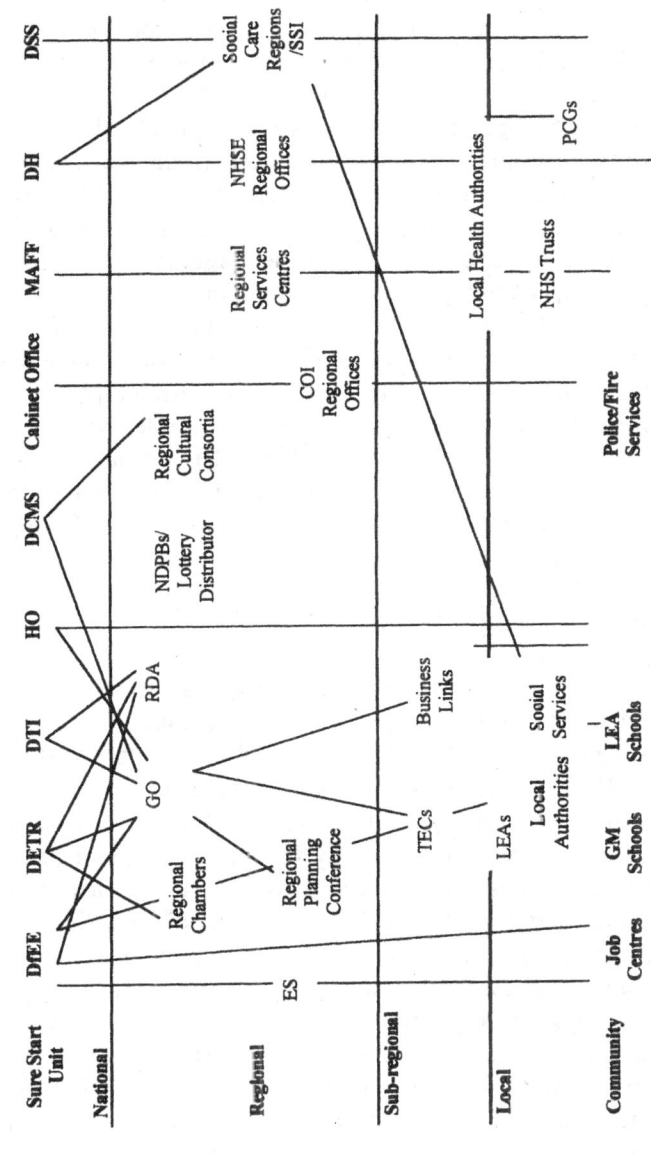

The current state of play on the production of RPG is outlined in Table 15.1. There may be up to five to six years between the start and finish of the cycle and the adoption of the guidance by the Secretary of State. A particular key step is the Public Examination. East Anglia has been the first region to go through the full process; SERPLAN also had a Public Examination in 1999, and following considerable debate the government has now pulled together responses to the Draft RPG for the SE (RPG9) (GOSE, 2000). Assessing Current Practice Assessment is difficult, and provisional. There is a complex array of new agencies, arrangements and policy frameworks, including the new Planning Policy Guidance note (PPG11) (DETR, 2000). It is still early days for a system in flux, and much institutional learning is taking place in RPG production, from early to later regions in the time schedule. There are also considerable variations in practice between regions. Nevertheless, it is possible to draw out some trends in, and evolving characteristics of, both RPG content (roles, issues and policies), and process (institutions, organisations and procedures). This also allows some assessment of variations in practice, and of 'value added' by the RPG process.

**Content: Roles, Issues and Policies**

The new RPG is intended to be wider in scope than formerly, and also to reflect regional needs and identity. A 'contents analysis' of recent RPG documents suggests some commonality of key issues, working to the agenda set by PPG11. These issues include land use, housing, economy and, to a lesser extent, transport; however, emphasis does vary between regions with for example, the economy having a dominant profile in the North East. Other issues, such as waste, water, energy, health and sport, are also important, but more second order considerations. Similarly, within issues, there are again some common areas. For example, under the development and land use issues, most regions are interested in balanced development, compact urban forms, and brownfield site development – but Green Belts are less of an issue in some regions, such as the East Midlands and East Anglia. Many issues (in particular housing, transport and waste) are more contentious in the South East than elsewhere, and the significant variation in issues may be less London versus the Rest of the South East (ROSE), and more the South East versus the Rest of the UK. Figures 15.2 and 15.3 briefly summarise RPG/draft RPG content, for the economy and for development and land use, across six regions.

**Table 15.1: Regional Planning Guidance Time Schedule**

| Region | 96 | 97 | 98 | 99 | 00 | 01 | 02 |
|---|---|---|---|---|---|---|---|
| East Anglia | Draft Strat. | → | → | Public Exam | RPG6 | | |
| South East | | | Draft Strat. → | Public Exam | Draft RPG 9 | | |
| South West | — | — | Draft Strat. | → | Public Exam | | |
| East Midlands | — | — | Draft Strat. | → | Public Exam | | |
| Yorkshire and Humberside | | | → | Draft Strat. | Public Exam | | |
| North | | → | → | Draft Strat. | Public Exam | | |
| North West | | | → | Draft Strat. | → | Public Exam | |
| West Midlands | | | | | → | Draft Strat. | Public Exam |
| London | | | | | Spatial Development Strategy to be prepared by Mayor for London | | |

Other current aims for RPG are for policy integration, within and between policy fields, and between policy levels (see Figure 15.4), and policy innovation. But there are problems. For example, within housing, a sustainable approach is to locate housing close to jobs, but this may conflict with the affordability of housing. Between policy fields there is often conflict between economic planning and land use planning, and securing integrated land use and transport planning can be difficult. Overall there is considerable 'messyness' of horizontal, and of vertical relations – including relationships with Government Offices, and with local authorities. To date there has also been only limited innovation, with policy lacking a radical edge, socially or environmentally – with, for example, little consideration of ideas such as 'environmental space' and 'livelihood strategies'. Migration and household formation, transport, and social exclusions, continue to remain difficult issues to handle for most regions.

**Figure 15.2: Regional Planning Guidance/Draft Guidance: Economic Comparisons**

| Economic Structure | East Anglia | East Midlands | South East | South West | North East | Yorkshire and Humberside |
|---|---|---|---|---|---|---|
| Development in Urban Areas | ❏ | ■ | ■ | ■ | ■ | ■ |
| Disadvantaged Priority Areas | ■ | ❏ | ■ | ❏ | ❏ | ❏ |
| Local Partnerships | O | O | ■ | O | ❏ | O |
| Infrastructure/Key Service Provision | ❏ | ❏ | ■ | ❏ | ■ | ■ |
| Quality/Increased Productivity of Labour | ■ | ❏ | ■ | ■ | ■ | ■ |
| Address Discrimination in Labour Market | ❏ | ❏ | ■ | ❏ | ❏ | ❏ |
| Knowledge and High Technology | ❏ | ❏ | ■ | ❏ | ❏ | ❏ |
| Clusters | O | ■ | ■ | ■ | ❏ | ■ |
| Diversification | ❏ | ■ | ■ | ■ | ❏ | ❏ |
| Research New Technologies | ■ | ■ | ❏ | ❏ | ❏ | ■ |
| Sectoral Weaknesses | ■ | ❏ | O | ❏ | ❏ | ❏ |
| Inward Investment | ■ | ❏ | ❏ | ❏ | ■ | ❏ |
| Address Areas of Economic Pressure | ■ | O | ■ | O | O | O |
| Rural Economic Diversification | ❏ | ■ | ❏ | ■ | ❏ | ■ |
| Major Sites | ❏ | ❏ | O | ■ | ❏ | ❏ |
| To Serve Local Communities | O | ❏ | ❏ | ■ | ■ | ❏ |
| Improve Image of the Area | O | O | O | O | ■ | O |
| Growth of Existing Enterprise | ❏ | ❏ | ❏ | ❏ | ■ | ❏ |
| SMEs | ❏ | ❏ | ❏ | ❏ | ■ | ■ |

■     Strongly stressed
❏     Included but not stressed as a principal issue
O     Not mentioned or only mentioned in passing

*Source:* Derived from content of RPGs/draft RPGs

**Figure 15.3: Regional Planning Guidance/Draft Guidance: Development and Land Use Comparisons**

| Development/ Land Issues | East Anglia | East Midlands | South East | South West | North East | Yorks and Humberside |
|---|---|---|---|---|---|---|
| Regionally Balanced Development | ■ | ■ | ■ | ■ | ■ | ■ |
| Support Peripheral Areas | ❑ | ■ | ❑ | ❑ | ❑ | ❑ |
| Urban Areas | ■ | ■ | ■ | ■ | ■ | ■ |
| Brownfield Sites | ■ | ■ | ■ | ❑ | ■ | ■ |
| Refurbishment | ❑ | ■ | ■ | ■ | ■ | ■ |
| City Centres | ❑ | ■ | ■ | ■ | ■ | ■ |
| Compact Urban Form | ■ | ■ | ■ | ■ | ■ | ■ |
| Close to Work, Shopping and Recreation Facilities | ■ | ❑ | ■ | ❑ | ❑ | ■ |
| Conservation of Historical Environments | ■ | ❑ | ❑ | ■ | ■ | ❑ |
| Protection of Rural Environments | ■ | ■ | ❑ | ■ | ■ | ■ |
| Urban Renaissance | ❑ | ❑ | ■ | ■ | ■ | ❑ |
| Greenbelt | ❑ | ❑ | ■ | ■ | ■ | ■ |
| Estuary Management | ○ | ○ | ■ | ○ | ○ | ○ |

■     Strongly stressed
❑     Included but not stressed as a principal issue
○     Not mentioned or only mentioned in passing

*Source:* Derived from content of RPGs/draft RPGs

**Figure 15.4: Problems of Integration**

A within policy fields

B between policy fields

C between policy levels

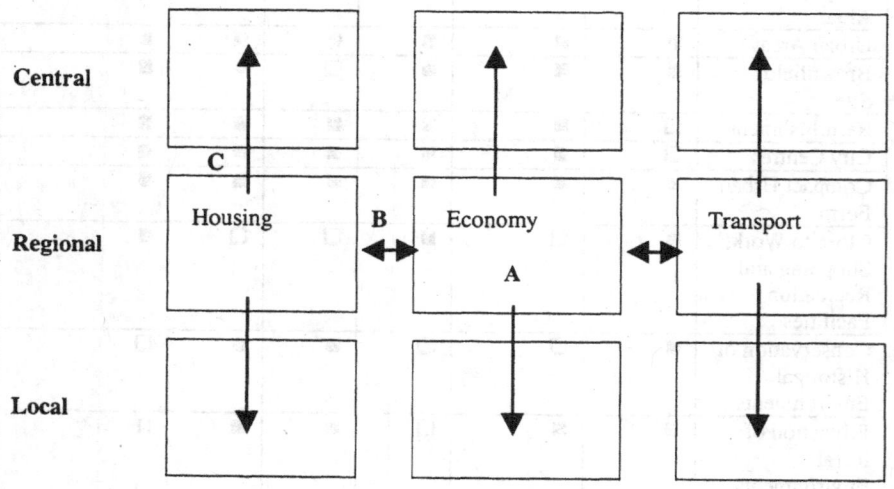

## Process: Institutions, Organisations and Procedures

The new regional planning process seeks inclusion, partnership and participation, but in practice this has proved difficult. The rapidly changing and complex institutional and policy environment of UK regional development and planning is clearly not yet 'joined up', as recognised in the Government's Performance and Innovation Unit's Report (PIU, 2000). Indeed in some areas it is very competitive, with RDAs pursuing vigorous economic agendas. For example, the East Midlands Development Agency has the aim of raising regional GDP per capita to the level of the top 20 EU regions.

Nevertheless, there is some interesting 'work in progress' on developing partnership approaches. As a minimum it includes the development of agency networks. But in some regions these have been evolving into more formal 'concordats'. In Yorkshire and Humberside the concordat is between the RDA and the Regional Assembly. In the North East, the concordat also includes the Government Office for the region, and has the following aims:

> One North East, the North East Regional Assembly and GO-NE will work together to help improve the economic performance of the North East Region to enhance the Region's environment and to improve the social well-being of all citizens within the Region.
> We will aim to achieve:
>
> - a common vision for improving the economic, environmental and social prospects for the Region's citizens;
> - complementary and mutually consistent strategies;
> - as far as possible within our respective powers, integrated implementation plans;
> - a shared understanding of what is being achieved, through monitoring;
> - a joint intention to keep strategies and plans under review with flexible mechanisms to enable changes in direction (One NorthEast, 2000).

Of particular interest is the Integrated Regional Strategy (IRS) approach being pioneered in the East Midlands Region (East Midlands Regional Assembly, 2000) where the relevant agencies have agreed a 'vision' for the Region, and are working together to integrate the various 'viewpoints' into a set of integrated regional strategies and policies (see Chapter 11).

Such innovations are having to confront a number of challenges. The evolving roles of some agencies, such as the RDAs and GOs, are unclear and there are problems of 'turf wars' about control. The GOs have a particularly ambiguous role, as both partner and controller, both 'in' the region and 'of' the region. The extent to which GOs are prepared 'to go native' for the region can raise tensions with other agencies. The ambiguity can perhaps be seen as a necessary element in the system, creating the space for political decision-making. Other challenges include limited resources. We have RPG on a shoestring budget. In-house teams are small, supplemented by some externally commissioned work. There is a need for more technical support, very usefully provided by local University – linked Regional Observatories in a few regions. There is also a need for continuity in the process, keeping a team together to monitor change and to speak for the guidance. Regional identity can also be an issue, especially in inappropriately drawn regions such as the South West and South East, and, in comparison with Scotland and Wales, the English regional level has weak civic arenas. Yet, not withstanding such concerns, there is probably more innovation taking place in the process than in the content of new RPG.

## Scope for Variations in Practice?

There is much commonality of agenda of content and of process, reflecting *inter alia* the considerable influence of national policies and guidance. But over the last two years significant differences in approach between regions are also beginning to emerge, especially in the process and institutional arrangements. Significant roles by leading actors, especially professional planners, can make a real difference. It is important to track and seek to explain this variability, and possible outcomes, over the next few years. In addition, it is important to check whether it all adds up, and adds value. Do the regional strategies add up to a national framework? Inter-regional issues are important, and most regional issues do not end at the artificial, and often unsatisfactory, DETR boundaries. Unfortunately, current evidence is not too strong on the inter-regional thinking in RPGs. In terms of added value it is probably too early to judge. However, if the process of RPG encourages better, and more open, working together of regional stakeholders, there may be more likelihood of support for the strategy which does emerge. As such the process of RPG may be as important as the product in terms of value added.

## Explaining Current Practice

Figures 15.5 and 15.6 seek to illustrate layers and dimensions of influence on current practice. The *layers* include State and Supra State, Regional Governance, Regional Planning and Local Governance. UK Government is currently creating opportunities for regional planning, but the complex rolling out of initiatives is also creating major headaches. The European Union (EU) provides a wider context, and a history of regional intervention. In its 'Europe of Regions' the EU has created major Structural Fund initiatives, and the history of joint working/bidding in some UK regions to produce Regional Development Plans to access EU funds provides useful 'institutional capacity' on which to build. Regional governance, already partly referred to in the previous section, provides a partially autonomous force of competing, and sometimes co-operating, agencies trying to influence the regional agenda; the role of business, via the RDAs, is notable. The regional planning community, through the skills of particular individuals, can also make a difference. Local governance is well embedded in the local government system and in local interest groups, although the planning function is becoming somewhat fragmented by tangential processes such as Local Agenda 21, and the forthcoming Community Plans and Local Strategies.

The *dimensions* of policy influence are partly encompassed in the concept of 'New Regionalism', as argued by Tomaney and Ward (2000). Four such dimensions are outlined in Figure 15.6. The economic dimension sees regions as now more exposed to global pressures than ever before. They are less protected by the State and have to look after themselves more. The environmental dimension incorporates environmental sustainability, recognises the importance of bio-regionalism, and identifies the region as the ideal arena for achieving 'development-environment' balance. The cultural dimension introduces the important element of identity. Some regions have a relatively clear sense of identity, and this is in some cases long-standing. But identity changes and the influence of media may be changing, particularly towards regional television. Finally there is the political dimension, and the movement towards more regional government. Whilst this is most clearly developed for Scotland, and to a lesser extent Wales, the introduction of the Regional Chambers/Assemblies for the English Regions is also significant. However, the level of demand for democratically elected English regional assemblies appears variable, perhaps giving some credence to Harvie's assertion (1991) that English regionalism is 'the dog that never barked'.

**Figure 15.5: Layers of Policy Influence**

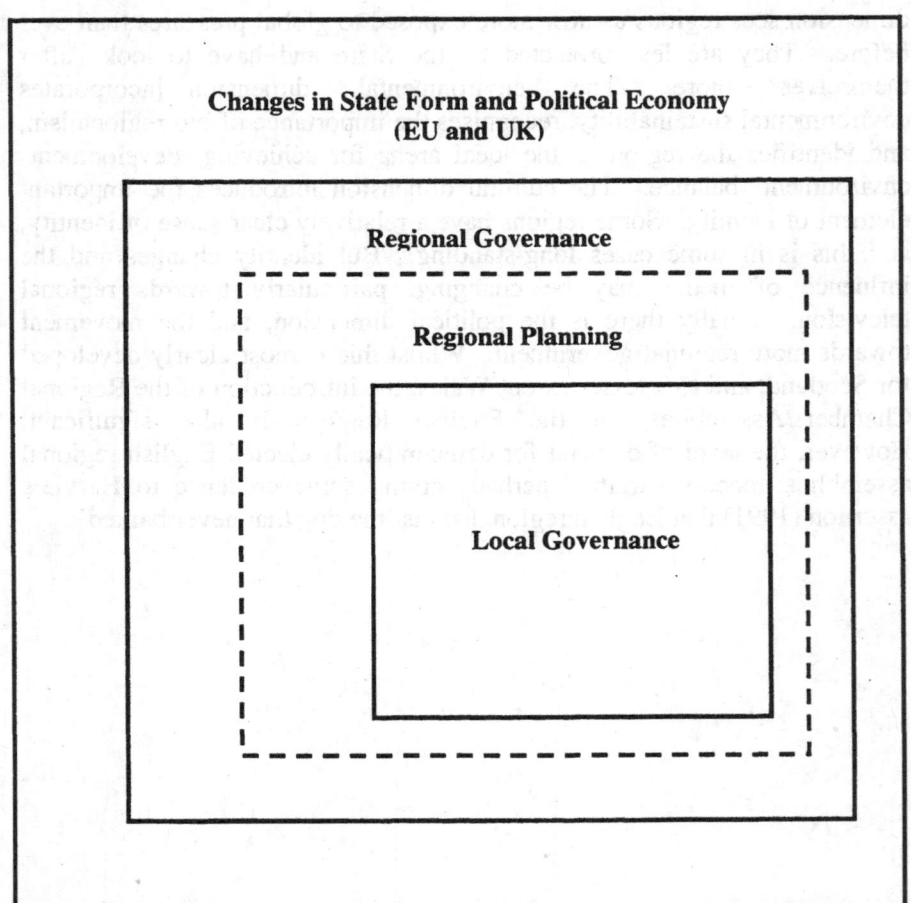

## Challenges to be Confronted

This interim appraisal of the evolving RPG and governance suggests that there are some important challenges to be confronted. The *constraints on RPG content* need to be relaxed, allowing a widening of scope and flexibility of content, to include more fully issues such as health, culture, and sport and recreation. *The limited integration and co-ordination* presents a further challenge. At a minimum there is a need for the better integration of economic development, and land use planning and transport. The production of a 'strategy of strategies' represented by the East Midlands Integrated Regional Strategy approach offers a promising way forward. The enhanced partnership approach illustrated by the North East and Yorkshire and Humberside concordats, is another useful step towards a more joined-up approach. Additionally, following the publication of the government's PIU report on regional agency fragmentation, referred to in Section 2, a Regional Co-ordination Unit (RCU) was established in central government. Its purpose was to fill the vacuum between Whitehall and the government regional offices. The RCU's 'Reaching Out Action Plan' (RCU, 2000) argues that government offices with responsibility for the regions should be more directly involved in policy-making. It also calls for better co-ordination of area-based initiatives to prevent policies and programmes overlapping, and for simplification of management structures. However, although such recognition of the need to co-ordinate the growing organisations of regional government is to be welcomed, there is still need for greater integration with the RDAs.

But *resource constraints* on the regional planning activity are apparent in most regions. Regional planning is currently operating on a shoestring, and a great deal of good will from various levels of government and relevant agencies. Such resourcing does not bode well for the future. There could be a lack of continuity with limited resources. There is probably also a need for better training on, for example, the nature of the evolving regional environment and approaches to multi-stakeholder planning. Such training should be for both members and offices, and should seek to reduce misunderstanding, fear and prejudice between the various tiers and agencies involved. Without such improvements there is a danger that the much better resourced RDAs will become the *de facto* regional planning agencies.

**Figure 15.6: Dimensions of Policy Influence**

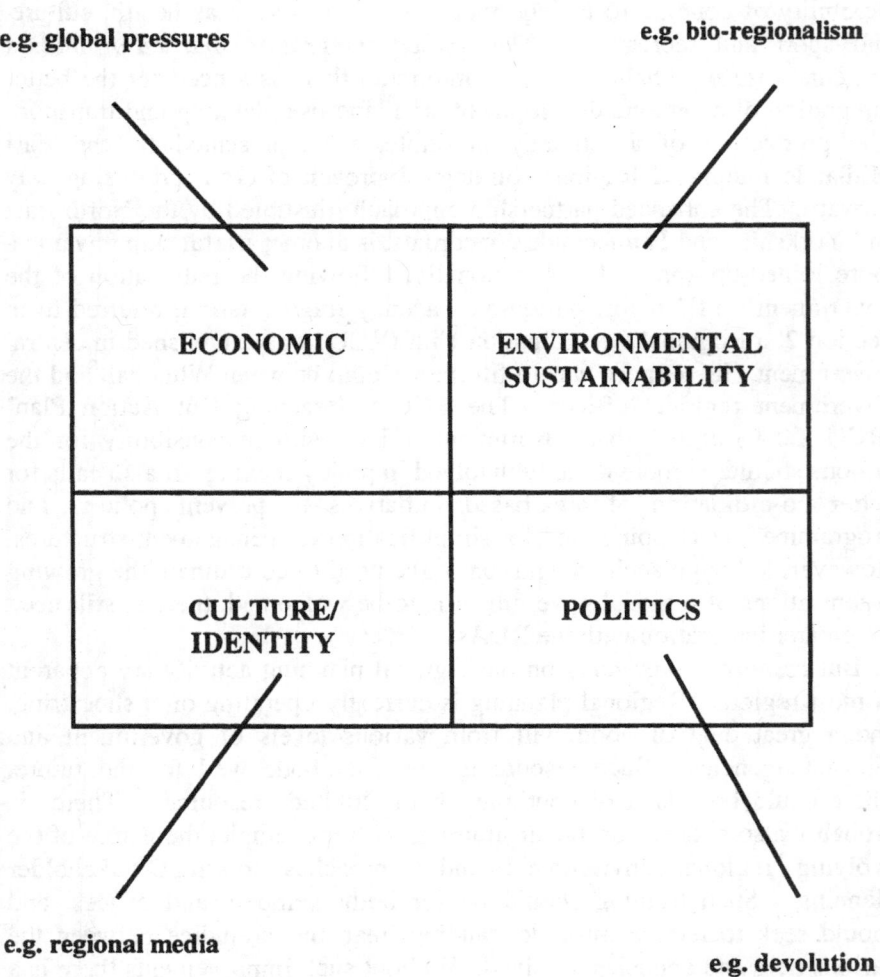

Other challenges include the *inward-looking nature* of RPG. There is a need to build inter-regional issues much more into the process. A simple innovation could be a more systematic approach to the inclusion of representatives from adjacent regions into each other's regional planning exercises. Co-ordination of approaches to common issues would also be helped by the production of RPGs on the same timescale – reducing the lags and leads in the system. More structural approaches may be provided by the growing influence of EU activity, through the European Spatial Development Perspective (CEC, 1999) in particular. In addition, bodies such as the Royal Town Planning Institute are encouraging the development of a UK national spatial strategy (RTPI, 2000).

Of course, it may be that some current inter-regional issues could be lessened with a different set of English regions. The *relevance of our current regional boundaries* is an issue of concern. Is the current scale of the geography of regions anywhere near right? There are arrangements for pan-regional working for the South East region, which may have 'knock-on' effects elsewhere. On the other hand, there is increasing recognition of the value of sub-regional working – for example in the 'Oxfordshire–Berkshire–Buckinghamshire' sub-region within the South East.

A further challenge to be confronted is the *democratic deficit* in the current activities. There is a need to improve public participation in the process, and this will require innovative approaches to convey the strategic issues involved. For the future, the case for directly elected English Regional Assemblies needs serious consideration, including the likely implications for other levels of government and planning.

**Interim Conclusions – Will the New Arrangements Work?**

There is scope for optimism. Regional planning and regional governance has a much higher profile than only a few years ago. Interesting new agencies and RPG procedures are emerging; there is also evidence of growing regional innovation. But it is too early to say whether this may yet be another false dawn. Indeed it is probably unrealistic to expect too much, especially in the way of technical achievements, when the agencies are still being set up and are seeking to come to terms with their environment and to clarify their roles. As the old Chinese proverb says 'May you live in exciting times' – the next few years promise to continue the state of excitement for regional planners.

# References

Breheny, M. and Hall, P. (1984), 'The strange death of strategic planning and the victory of the know-nothing school', *Built Environment*, vol. 10 (2), pp. 95–9.

CEC (1999), *European Spatial Development Perspective*, Committee on Spatial Development, Potsdam.

Department of the Environment, Transport and the Regions (DETR) (1998), *The Future of Regional Planning Guidance*, HMSO, London.

Department of the Environment, Transport and the Regions (DETR) (1999), *PPG Note 11: Regional Planning – Public Consultation Draft*, DETR London.

Department of the Environment, Transport and the Regions (DETR) (2000), *PPG Note 11: Regional Planning*, HMSO, Norwich.

East Midlands Regional Assembly (2000), *Integrated Regional Strategy: Draft Summary for Consultation*, EMRA, Melton Mowbray.

East Midlands Regional Local Government Association (EMRLGA) (1998), *Regional Guidance for the Spatial Development of the East Midlands*, EMRLGA.

Government Office for the South East (GOSE) (2000), *Draft Revised Regional Planning Guidance for the South East (RPG9)*, HMSO, London.

Harvie, C. (1991). English regionalism: the dog that never barked, in B. Crick (ed.) *National Identities: the Constitution of the UK*, Blackwell, Oxford.

Martins, M.R. (1986), *An Organisational Approach to Regional Planning*, Gower, Aldershot.

One NorthEast, North East Regional Assembly, and Government Office for the North East (2000), *Concordat between One NorthEast, the North East Regional Assembly and the Government Office for the North East*, One NorthEast, Newcastle.

Performance and Innovation Unit (PIU) Cabinet Office (2000), *Reaching Out: the Role of Central Government at Regional and Local Level*, PIU, Cabinet Office, London.

Regional Coordinating Unit, UK Government Cabinet Office (2000), *Reaching Out: Action Plan*, Cabinet Office, London.

Royal Town Planning Institute (2000), *The UK Spatial Planning Framework: A Discussion*, RTPI, London.

SEEDA (1999), *Building a World Class Region*, SEEDA, Guildford.

Tomaney, J. and Ward, N. (2000), England and the 'New Regionalism', *Regional Studies*, vol. 34 (5), pp. 471–78.

Wannop, A. (1995), *The Regional Imperative: Regional Planning and Governance in Britain, Europe and the USA*, Jessica Kingsley, London.